THE MONEY MAKING POWER OF

GOLD

Timeless Strategies for Profits

CHRISTOPH EIBL

ISBN 1-59280-296-6

Printed in the United States of America.

1 2 3 4 5 6 7 8 9 0

CONTENTS

GOLD

FOREWORD

Mankind has always been fascinated by gold. The noblest of metals and metal of the nobles has seen empires rise and fall, has persisted through wars and financial crises and is, to this very day, the only currency to have maintained its global buying power for millennia. At a time of widespread worry about the sustainability of our current monetary system, the development of the gold price has once again attracted eager attention. But very few initiates know how this price is determined. Where does the heart of the gold market lie and which gold prices are published? How does the ore recovered by the gold mines acquire its final bar shape? Where are the centers of the jewelry industry and how is the physical gold distributed there? Who are the gold market's major players and what are their trading customs? How are the Asian and Arab gold markets developing? Much to his merit, Christoph Eibl addresses these questions in this book. He thus closes an important knowledge gap in the literature on gold. Private investors and asset managers alike are provided an easy-to-follow introduction to the multifaceted world of the international gold trade. And as a reference on the gold market, the book in hand will be just as valuable to analysts and finance professionals.

The author assumes the perspective of a professional gold trader to offer a sophisticated overview of various gold products, ranging from bars through gold lending to gold futures. The reader finds out about the sums shifted in the gold market and discovers to what extent this precious yellow metal can diversify even very large equity portfolios. Given the substantial economic imbalances in this world of ours, the run for gold and raw materials seems bound to continue. Better start reading immediately!

—Markus Mezger Stuttgart, February 21, 2005
Senior Economist, Baden-Württembergische Bank AG

GOLD

PREFACE

There has to be a reason to write a book and the work in hand is no exception. I would therefore like to answer the following question for you: Why write a book about the gold trade?

The gold market is one of the oldest financial markets in existence and was once widely regarded as a dangerous place, mostly during the 1970s. Yet a price hike of $35 USD per ounce to well over $850 USD shows just how active and interesting it was at the time. But when it finally arrived in the 21st century, the gold market was something of a has-been and rarely found mention in the media. The professional trade's know-how had noticeably waned during the 1980s and 1990s. The result was a market with elderly gentlemen traders and little dynamism, a type of lethargy having set in.

All that is a thing of the past. Today, the situation couldn't be more different: Gold is hot. Gold is more volatile than ever (not quite as much as in the 1970s, but almost as dynamic), and it is perhaps more interesting now than it has ever been before. Today's agenda is set by expansions of production, a potential bond crash and budget deficits around the globe. That peculiar fascination the gold market traditionally used to engender has returned. There might be some very good reasons to look at gold a little more closely, after all. But where is one to find relevant information?

One way, if not the most obvious, would be to gain knowledge from literature. But my search quickly drew to a close. Hardly any specialist literature exists in German and even the odd volume located here and there usually dates from the 1970s or 1980s. One might be tempted to think that the search for material, if it was extended to the Anglophone world and hence the gold price fixing's country of origin, would be more rewarding, especially considering that the international center of physical gold trading is located in London, even today. But once again, back to the drawing board. The literature I found is at least a decade old and rather meager, with the exception of a few mentionable, individual works, such as Timothy Green's (1993). What I thus discovered is an enormous gap in the literature on gold; it is this very gap this book intends to address.

But let me first make some general observations on the book's subject matter. The gold market differs from other classic, bank-operated financial markets in a number of ways, one of the most important being its marked opacity, which can sometimes make it appear mysterious and hard to fathom. Gold fixing, spot transactions and central bank sales take place bilaterally, without open trading and without being reported. This means that trading volumes are as hard to identify as the market participants.

Only if one were to enter this world for professional reasons one could slowly grow into these structures. One would make the acquaintance of some of the central players in the London Bullion Market Association, one would find out who generates the largest transactions and one would even gain insight into how some of these trades function at all. The learning curve looks daunting, and I sometimes even asked myself whether I would one day be able to understand every detail of this market's enormous depth. It would probably take decades of intensive gold trading, if anything.

I shall still make an attempt to show my readers just how the gold market pulsates and survives, basing it on my professional experience; my visits to South African and South American mines as well as congress attendances in London, North America and Russia; countless long hours spent on the trading floor; and frequent, seemingly endless business trips to Brazil, Switzerland and the Middle East. Whether you are reading this book as a private or institutional investor, a trader to-be, a jeweler or a scientist, I am convinced that you will find a lot of new, interesting and useful information in it.

INTRODUCTION

It is impossible to understand the gold market and its immanent structures without looking at its past, at least in basic terms. It might be tempting to start with the gold of the Egyptians, Romans and Scythians. Their minting artistry and jewelry skills are impressive, even in view of today's possibilities. The very first coin mentioned in history came from the Lydian Kingdom. The first standardized gold coin of the Roman Empire, which also served as a name patron for the chemical symbol for gold, "Au", was the aureus. Aurora was the Greek goddess of dawn and Aurum would thus become the Latin word for gleaming gold. The story of gold with all the wars and conquests of various periods is one of great interest and high tension. But in terms of today's trading structures, the real history of gold started much later: March 15, 1968.

This day brought an end to the fixed-price gold policy that had been valid for decades. In the years before 1968, gold was always traded between central banks and on the inter-bank market at a fixed price of $35 USD per ounce. A fixed trading price naturally limits the possibilities insofar as fundamental imbalances of supply and demand may not be accounted for in the pricing. The gold price had, therefore, during the period leading up to 1968, reflected an unrealistic market view. Now a variable gold quotation, the two-tier gold market, was introduced. This meant that the inter-bank market and the central banks simply continued to trade gold at a fixed price of $35 USD, while the remaining market did so on a free price basis.

When the USA gave up its gold standard in 1971, the Bretton-Woods treaty became void. The market would never return to its old self after that. Instead, it developed individual structures and its very own dynamism. An international global market very quickly established itself, and gold could now be traded 24 hours a day, up to six days a week (a six-day week being customary in Hong Kong, where trading continues on Saturdays). The true significance of the decision to abandon the market to its inner dynamics (some high finance personalities called it a "liberation coup" for the gold price) may be gleaned from the ensuing price development. The new gold price was defined by heightened volatility. In January 1980, the troy ounce reached a new price peak at $850 USD per ounce. But this was followed by a lengthy downtrend lasting until 1999 with a low of only $250 USD per ounce.

The market that developed from this background radically differed from all others. Numerous exchanges started for all kinds of gold products. Gold was now traded in kilograms, Tael, Tola, ounces and various finesses like .9999, .9950—London Good Delivery status. Options and futures in gold came into being. Although each market had individually adapted to its local contestants, one could still be sure of round-the-clock trading. This was particularly important because gold accounts could henceforth be transferred globally. Hong Kong traders could pass the accounts on to London in late afternoon and London would pass them to New York in the afternoon again. But even while the gold market was no longer subject to a fixed exchange rate, it had by no means been liberated from statutory regulations entirely. Some countries outlawed private gold ownership (which is still the case in South Africa today). Elsewhere, the commercial non-bank trade by was prohibited or gold purchases were subjected to a special tax, as was the case in Germany until 1993. The result was a flourishing gold smuggle that was able to supply the precious metal virtually anywhere in the world. In his book "The World of Gold" (1993), Timothy Green vividly describes how even within India itself gold used to be continuously smuggled from one state district to the next due to varying levies.

But when we speak of the gold trade market, we primarily mean the physical or derivative trade between institutions such as banks, hedge funds or mines. This book's focus, too, lies on the inter-bank trade and the trade between banks and clients. And to further the reader's contextual understanding, the whole procedure of gold production and processing shall be explained: from the ore to the molding of bars up to the manufacturing of jewelry.

Anyone attempting to track the passage of the gold ore into the financial markets will find himself confronted by a dense and multi-layered network of connections between banks and mining companies: so-called bullion banks, mostly operating from London, Zurich or Sydney, finance drilling, exploration and expansion projects undertaken by mining companies. The companies thus financed in turn use the banks for their coverage and trade transactions. This results in a double yield for the banks that compensate for initial credit risks. The trades usually include hedging strategies such as forward, futures and option transactions.

GOLD

The actual process starts in the mines. The mines initially produce the so-called Doré bars that still contain other ore components like silver or copper besides the gold. These bars are produced from the ore immediately after extraction, sometimes directly by the mine itself, if the operation is large enough to warrant such an effort. There is no need for a first-class refinery to be present locally to create Doré bars. These "raw goods" are then transported to internationally renowned refineries like WC Heraeus in Hanau, Germany, or Johnson Matthey in London, England, which proceed to refine the Doré bars and cast them into high-quality bars (24 carat or .9999) on site.

But even before the Doré bars leave the mine, the mining company, insofar as it handles the gold distribution itself, has to decide the market in which the future material is to be sold. The reason being that most marketing centers require products, qualities and references that vary from one location to the other. This means that bar sizes, units of weight, the quality and, most of all, the respective refinery's status need to be considered. In some countries only bars from accredited refineries may be traded. In Germany, bars from German refineries are always accepted, while other brands are often traded at a discount. Whereas the member circles of the London Bullion Market Association (LBMA) will only accept bar qualities holding the official Good Delivery status.

The professional market trades and prices gold on the basis of 400-ounce (12.5 kilogram), London Good Delivery bars. The LBMA sets very high standards for bar producers wishing to supply these bars. The association has to this end established special statutes setting out the conditions for the bars and their producers. Banks, central banks and asset managers therefore often hold their stocks in Good Delivery bars to guarantee their international fungibility. Outside the gold investment market, industry frequently has a demand for granulate. Granulate or shot consists of small gold pellets that can be converted into a variety of shapes by relatively simple means. Shot is the format complying with standard demand at classic gold industry locations such as Pforzheim, Arezzo or Hanoi.

A closer inspection of the procedure will show that banks play a central role at virtually every stage. More on this in chapter 3.

Golden Qualities

The word gold is Indo-Germanic and derives from ghel which could be loosely translated as gleaming, shimmering or shiny. Gold has the chemical symbol "Au" and ranks as the least soluble of all precious metals; only aqua regia (a mixture of hydrochloric and nitric acid) can successfully attack this single isotope element. Gold has a very high density at 19.3 grams per cubic centimeter and a relatively high melting point at 1,337 degrees Celsius.

Gold is found virtually anywhere in the world. The best known extraction areas, which provide around 50 percent of all the gold ore mined, are located in North America, South Africa and Australia. Up to now, about two-thirds of all known gold supplies are thought to have been exploited—the mathematical equivalent of a cube with 18-meter sides. Today, annual extraction weighs in at 2,400 tons; it was only one percent of that 120 years ago. This increase has been facilitated by optimized extraction techniques, for example, in underground mining at a depth of over three kilometers below ground level, as well as new extraction processes involving various chemicals.

The veins of gold we all know from the movies, with their long and visible stretches of pure gold, are a product of the imagination—they just don't exist. If you are a very lucky mine operator you might here and there come across an underground section containing 15 grams or more gold per ton of ore. But your average mine operator has to make do with much less than that. Extraction techniques address this fact: Modern heap leaching processes permit the viable extraction of gold from ore containing fewer than two grams of the precious metal per ton.

Gold is frequently used in industry, most of all because it is extremely resistant to corrosion. This feature and gold's relative malleability still make it the primary stock-in-trade in dentistry, where it is used as a filling or substitute material for damaged teeth. The electronics industry also uses gold because of its high resistance to corrosion and optimal processability. The tiny wires connecting computer chips to other components are manufactured from highly purified fine gold—with only one gram of the precious metal easily extended to a wire length of several kilometers.

THE RANGE OF GOLD

THE RANGE OF GOLD
MARKET PRODUCTS

-I-

In this chapter I shall try to explain just how complex and far-reaching the gold market can be. After the introduction of various commercial products and financial instruments I shall take a closer look at their significance and structure.

At first sight, gold trading looks as easy as pie. You buy the gold, have it delivered, and sell it after a price hike. Reality isn't anywhere near as simple. As a matter of fact, there are so many details and peculiarities lurking in this apparently basic operation that a novice would find it very difficult to master such an apparently simple trade, at least on a professional level.

One special feature that is characteristic of all raw-material markets—and therefore the gold market—and simultaneously serves to sharply delineate them from stock markets is their physical component, the term physical meaning an effective delivery of the goods. At this point, it becomes apparent that an additional process has entered the playing field. Logistics, insurance and storage play important roles, apart from trading in terms of bid and asked quotations, i.e., buying and selling prices. You can't simply load a truck with gold and transport the precious metal from South Africa to Zurich on a pallet without further ado. The metal needs also to be stored with the appropriate insurance coverage. And one should also not forget about the bank vault's structural engineering, as gold has a very high density, which means that even a small volume may, due to its weight, result in massive storage problems.

Another unusual feature of raw-material markets is the commodate loans in the futures market. Situations of surplus supply or demand affect the lending market for physical gold via various factors displaying typical "raw-material" traits. Delivery bottlenecks, warehouse gluts, production downtimes, mine closures or new gold ore discoveries are only some examples. These and further unusual features of the gold trade will be examined more closely in the following chapter.

PRODUCTS OF THE INDUSTRIAL AND INTER-BANK SECTOR

No two customers are alike, and gold therefore doesn't automatically equal gold. Huge differences exist between one trader and the next. Volume is only one client classification indicator, product choice being another. Based on this realization one can divide the gold market into industrial products, products for the inter-bank market, and products for private or investment clients.

Industrial clients usually assume a singular role on either side of the market, i.e., they are either pure buyers (as in the jewelry business) or pure sellers (as in the mining industry). Industrial clients in the mining sector are marketeers of gold and perform their sales either directly or via various financial products. Mining enterprises relatively often rely on hedging products, for instance. The term "hedging" describes the safeguarding of future production against possible price fluctuations by using the appropriate financial products. The market approach of industrial clients as well as the central banks is usually quite cautious, because its primary intention is often to not upset it.

At the industrial clientele's opposite end of the spectrum are businesses called natural takers. The classic example for this would be the jewelry industry. Jewelry production requires gold granules that may be forward financed by banks more cost-effectively, compared to an immediate purchase. The term of this type of forward financing is usually 30 to 90 days. During this time period, manufacturers turn the gold pellets into finished products, which they sell. They then use the proceeds to meet their liabilities as specified in the future contract.

THE PHYSICAL GOLD MARKET

A principal distinction is made between trades requiring additional transactions that are typical for raw materials and trades of a purely financial nature (i.e. trades lacking the physical component). Physical gold comes in all shapes and sizes and is traded accordingly.

On the industrial side, gold wire, gold plate, gold bands and gold granules are traded and sold. These types of gold usually come from refineries and are traded in non-bank markets. The gold has been processed and commands higher prices due to the higher investment in terms of both labor and time. The following chapter is based on gold in the shape of bars that is customary in industry and the inter-bank trade.

THE CASH MARKET

As a gold investor you quickly come across the term bullion—meaning bar-shaped gold that is therefore marketable institutionally. Two other commonly used terms are ingot and bar. "Ingot" usually denotes small bars (i.e. ounce and gram bars), whereas "bar" mostly describes the classic London gold bar, as well as bars in other shapes and sizes. Depending on where you are in the world you'll encounter gold bars in a bewildering number of variations.

In continental Europe, the standard is the kilogram bar with a fineness of .9999 (or "four nines"). The Hong Kong market is based on the so-called Tael, a donut-shaped bar. And Indians much prefer the Ten Tola bar, usually abbreviated to T.T. bar. Global trade is still possible, quite despite the fact that almost every single nation has its own preferences, but sadly, only under the condition that globally active traders and arbitragers possess the specific knowledge required for local market conditions. It is hence of paramount importance to have permanent access to the relevant know-how concerning delivery costs, shaping costs, storage locations and their corresponding conversion modalities.

We shall first take a closer look at the spot trade or cash trade, as it is also known. When traders deal in physical gold in the inter-bank market this is principally referred to as a spot trade. Spot (spot next) designates the fixing of the availability date, i.e. the valuation of the gold, or the currency payment (usually USD), respectively. In the absence of other agreements gold is traded with an availability date of two days.

Tola Bar, Tael Bar, Kilogram Bar
Source: *Gold Refiners & Bars Worldwide*

There are some similarities with the currency trade. "Gold deliveries" are hence carried out over two days. Yet in real life there is no physical transfer, unless indicated otherwise. In the inter-bank trade, gold transfers are only posted electronically. The gold thus transferred is called unallocated gold; that is, gold that is stored in collective deposits with no special subdivisions representing specific owners. In terms of storage, the precious metal is divided into unallocated and allocated gold, i.e. collective and individual deposits. This distinction is very important for two reasons. Gold in individual storage in allocated accounts is held by the trader on behalf of a certain client. Individual bars can be identified on a weight list by their bar number, total weight, fineness and the refinery that shaped them. The metal is also physically separated from other bars in the vaults. This procedure is required so clients can, even if the trader goes bankrupt, directly claim their gold without it being included in the general insolvency assets.

The most common form of trade storage is in unallocated accounts. It is simpler and cheaper, also because these accounts can range into the debit side, similar to a cheque account. This is, of course, not possible with allocated storage. Unallocated storage thus enables gold transfers that closely resemble transfers of credit money. The significant factors are the right of ownership and a possible debtor-creditor relationship. A gold position in an unallocated account entitles the client to the ownership of an accordingly sized gold stock, but not to special or separate bars. All relevant positions are covered for the client by the trader's general gold holdings. There is therefore a risk of trader non-payment for clients who hold their gold in unallocated accounts, but in the opposite direction (i.e. with a negative balance), the trader also incurs a credit risk with his customers.

In all of these cases the trader could principally be a straightforward trading company, but in reality this function is usually performed by banks.

It hence becomes apparent that allocated gold is, due to the higher outlay, more expensive than unallocated gold. Another decisive factor is the client's or trader's, whichever you prefer, borrowing power. One of the results of this situation is that gold market participants trade among themselves, because they already know each other and have therefore been able to formulate mutually binding criteria that need to be met to enter their circle. The London gold market, for instance, features a trader's association in the shape of the London Bullion Market Association, which has served to de-

fine such mutual trading practices. Quite similar associations also
exist in Dubai, India, Hong Kong or Shanghai. But the individual
markets will be analyzed in greater detail in the following chapters.

Because the physical gold is not being traded institutionally via
any important exchange and can therefore only change hands over
the counter (OTC), some further components arise that also need
consideration.

Each bar carries an embossment of the gold's fineness. The highest
gold content is described by the figure .9999, the number express-
ing the gold content per 10,000 parts.

Vault of the Bank of England
Source: *Bank of England*

For the London inter-bank trade, a purity of .9950 is the minimum
requirement. German and Swiss refineries usually produce .9999
bars. The Hong Kong market customarily trades in the fineness
.9999, while bars for the customer segment usually have a purity
of .9900. The Chinese Great Wall Refinery produces bars with a
gold content of .9590, whereas the Singapore trade resembles the
London trade. The reasons why different markets trade different
finenesses are often rooted in local traditions to do with gold jewelry. The following table shows the most important fineness in the
relevant country's jewelry:

Gold in Prozent	Content Purity	Carat Number	Typical Markets
99.99	.9999	24	Hong Kong
99.00	.9900	23	Hong Kong (Chuk- Kam bars)
95.90	.9590	23	China (traditional)
91.60	.9160	22	India, Pakistan, Turkey (traditional)
87.50	.8750	21	Middle East, Thailand
75.00	.7500	18	Europe
58.50	.5850	14	USA, Germany
41.66	.4160	10	USA
37.50	.3750	9	Great Britain
33.30	.3330	8	Germany (minimum gold content)

Gold Purities and Their Markets
Source: *World Gold Council, author's own research*

Apart from the purity of the gold, the size and type of the bars are also important factors. Certain bar sizes (weight classes) are dominant in certain trading circles or trade associations, respectively. The London Bullion Association trades in bars with a minimum size of 400 ounces (the approximate equivalent of 12.5 kilograms). In India, the customary size is the T.T. (Ten Tola) bar. The Chinese use Tael bars in the inter-bank and client trade. Even this short comparison should suffice to demonstrate how important bar size and weight class can be in addition to the metal's purity.

To guarantee a truly global trade, bar sizes and gold purities need to be adapted to each relevant market. In professional circles these procedures are known as quality, size and loco swaps. As we shall see, and as the name suggests, these transactions are various types of exchanges.

In a quality swap, gold of a lower quality (lower purity), may be up-valued, as the name implies. The same applies to gold of a higher purity being swapped for lower quality gold. The size swap deals with bar sizes. Kilogram bars are swapped for ounce bars or Tola bars for Tael bars, for instance. Let us examine a brief example (for a more detailed explanation, please see Chapter 5, "The Most Important Trading Strategies in the Gold Market"). The official listing, Loco London, refers to London Good Delivery bars with a fineness of .9950. Let's say the selling price is $450 USD per ounce of gold. If buyers would like to find out the price of the fineness .9999, their exact calculation would therefore look like this: 450/9950 = 0.045. The result needs to be multiplied by four (because of the purity of .9999). The additional expenditure in this case would therefore be $0.18 USD per ounce.

This example illustrates the exact calculation concerning fineness. In reality, the pricing is completely different, because if such a swap was performed practically, meaning that the gold was to be tried, the reshaping or refining costs incurred would have to be added to the price. And because bullion banks are usually amply provided with gold positions of numerous different sizes and purities, this type of refining is only very rarely needed.

The term "size swap" refers to the transformation of Tola bars into kilogram bars, for example. Swaps of this type are frequently called to transform large London 400-ounce bars into smaller one-kilogram investment bars.

Loco swaps play a role in physical distribution and arbitrage. If customers in Dubai intend, for example, to purchase gold via the London gold fixing, they will receive it at a Loco London price. "Loco" in this case refers to the place of origin.

London Good Delivery Bar
Source: *Rand Refinery*

The Loco London price is an international trade and reference price. Some marketing centers, such as Hong Kong or Singapore, trade Loco London gold in addition to their own markets. If the aforementioned Dubai clients wish to store their freshly purchased London gold in Dubai, they can swap it. All they need to do this is a bank entertaining depots or consignment depots, respectively, in both locations. This bank effects an internal posting so its customers can obtain their gold locally in Dubai. This, of course, means the gold isn't actually and physically transferred. The prices for this procedure can vary a lot and depend on the following factors:

1. A local shortage of the material, for instance during the Diwali season in India. (See the section on that country.)
2. The unavailability of the material in local depots.
3. Whether costs are lowered by rising total volumes.
4. The unavailability of consignment companies.
5. Whether insurance premiums fluctuate widely.

Some readers may ask why the USD gold price of the London trade has such an important role to play internationally, at least in the physical segment. The historical development of these market structures is certainly one of the reasons, but another is supplied by the very high standards the active market participants have set themselves here. Although we shall take a closer look at their association in chapter 4, this seems to be the appropriate place for consideration of the London conditions concerning delivery quality.

Entrants into the London market accept gold in bars carrying the status London Good Delivery. The association has, in its bullion commission, created a list of assayers and refiners whose products have been tested and declared suitable. These companies meet the conditions regarding bar size and quality and can therefore be called upon to deliver for OTC commitments entered into in bilateral trades. This is the so-called Loco London gold, as we have seen. The bars have to be hallmarked accordingly and are required to meet the following characteristics:

1. The refinery's name and hallmark must be featured.
2. The quality (for instance .9999) needs to be embossed in four places.
3. The weight needs to correspond to a customary unit of measure.

4. They must contain a serial number which is recorded in a
   bar list at the refinery.

Loco London refers to the fact that the local delivery needs to be
performed at an accepted bullion bank. In INCO terms this would
be the equivalent of CIF London (cost, insurance, freight), which is
a professional way of saying the delivery and insurance costs up to
London are included.

Physical gold is only transported if absolutely necessary. There
are several reasons for this. Consignment costs are very high. To
move large amounts, a strategic consignment plan may be required,
as even a small amount in terms of volume can be expensive. It is
very important in this respect to collaborate closely with experienced carriers of valuables. Past experience has proven Brinks,
Securicor and Via Mat to be particularly reliable companies. And
if a gold delivery is to be transported free of risk, the insurance is
another cost factor.

The gold market's logistical processes are just as complex as the
metal's storage and safekeeping requirements. If the amount of
gold to be stored exceeds a certain volume, the action taken depends entirely on the safe room's load-bearing statics. It is for this
very same reason that an engineer needs to be consulted whenever
gold is added to a safe room's contents. Due to the material's enormous density, a cube with a side length of a mere 37 centimeters
weighs approximately one ton. Vault storage is usually performed in
pallet style on metal shelves that are distributed throughout the room.

After having somewhat familiarized ourselves with the trade's
physical components, we can now move on to the second question:
How do banks, central banks and mining companies actually trade
with each other?

The customary trading and communication platforms for this are
Reuters Dealing, which is run by the information service Reuters,
and the trade platform EBS, which is run by the Eponymous
Company. Both systems were traditionally used for currency transactions and this paved their way into the precious metals trade.
Reuters Dealing could be compared to a chat room where banks
establish contact with each other, using an online platform that's

similar to the Internet, and thus create a situation where they can discuss their deals via text conversion. Transactions are executed in written form. All communications are documented so that possible queries arising later on can be resolved. All market participants have individual abbreviations—usually consisting of four letters—resembling their ID codes, which can be used to contact them. The product is a market standard. All traders who are seriously engaged in the gold trade use this system.

EBS, however, is more of a pure currency product, which has also been quoting gold and silver prices for a number of years now. On EBS, market participants continuously quote bid and ask prices. Although gold is something of a sideline on EBS, the service is often used as a market price indicator or as a price setter in market periods lacking market makers. EBS has a distinct disadvantage compared to Reuters Dealing as the service does not feature a communication platform and only supplies liquidity instead.

In recent times, market makers increasingly supply proprietary trading platforms where they continuously quote prices for their clients. Their function is comparable to that of EBS, although with the added disadvantage that one is reduced to trading with one and the same trading partner all of the time.

In addition to these electronic trading options, there is also the gold trade via the telephone. Brokers, in particular, maintain leased lines to the major firms and are thus able to continuously distribute prices. They play a major role in the precious metals trade because they significantly increase the market's price transparency. By supplying clients and banks with market-maker prices all day long, a "real price" can be deduced more easily, especially because the price indications supplied to the financial world via the major data providers Reuters and Bloomberg often significantly differ from the market prices actually being traded.

Certain trade volumes are customary in this OTC market. Gold spot is quoted by market makers for up to 5,000 ounces. Whoever wishes to inquire about prices for larger volumes first needs to show this interest. The minimum quoted is usually 2,000 ounces. One also speaks of price inquiries for five bars, as 2,000 ounces is the equal of five London Good Delivery bars of 400 ounces each.

GOLD

Let me try to illustrate a gold trade transaction more clearly with the following example: Let's say the gold price ranges between $423.50 and $424.25 USD per ounce (this is how the current quote would be distributed at various data providers). The following conversation ensues:

Trader: "Spot gold, please" (inquiry for the gold price)
Broker: "80/20, the figure 4" (price quote)
Trader: "Five Bars, mine at 20" (desire to sell or buy)
Broker: "I'll sell you 2,000 at $424.20" (confirmation)

In this example, the client has just bought 2,000 ounces via the broker. "The figure 4" describes the last digit to the left of the decimal and in this case the number 424.

The continuous intraday trade in the gold market is augmented by the London gold fixing. This is held twice a day, at 10:30 am and at 15:00 pm GMT, respectively, and is of international importance. Other countries perform their own daily fixing as well, which will be explained in greater detail in chapter 4. The process of fixing is a permanent peculiarity of the gold market, although it has only really established itself in the London market. I shall describe the process in general terms and using excerpts from the real London fixing below.

But first of all, a comment seems to be called for. If I had written this book a year ago, the following description of the fixing process could have been left just as it is. But the gold market is constantly changing, and in March 2004, the highly traditional bullion establishment, NM Rothschild, announced its immediate withdrawal from the gold fixing and the gold market. This announcement represented not only a sad loss for the market, which probably lost its most experienced market member, but it also meant that the oldest financial instrument (i.e. the gold fixing) had to undergo a complete revamp. In the following, I shall describe the type of fixing process as it used to be for many decades. All the changes performed on this solely concern the fixing's efficiency. The fixing members no longer meet personally, the fixing chairman is now rotated annually, the fixing takes place via the telephone and tradition has all but gone overboard; however, the function, purpose and meaning of the London gold fixing remain exactly the same.

London Gold Fixing in the Rothchild Offices
Source: *N M Rothchild*

A fixing is a trade process aimed at trading as many transactions as possible at a fixed price. Fixing has the following advantages:

1. A uniform price all the traders can base their transactions on.
2. An internationally accepted (in local centers only for the respective market) and published price.
3. Large trade volumes can be processed.
4. No spread is due in fixing, as would be sometimes the case in the course of normal trading. Instead, transactions are executed via a discount and a premium.
5. Transactions may be executed in USD or other currencies, if desired.
6. Fixing standards ensure the quality of market participants.
7. The fixing price represents the gold price because the largest volumes are traded at the uniform price. This price is less volatile than the price in intraday trading.
8. High market liquidity due to significant order volumes (central banks/ production).
9. Absolute anonymity.

Shortly before the fixing starts, the intraday trade seems to freeze up. The entire gold community waits for the fixing to commence.

GOLD

It can become difficult, if not altogether impossible, to obtain an acceptable spot price during the fixing. Every fixing is presided over by a fixing chairman, who, in the London gold trade, has been traditionally provided by that keeper of traditions, NM Rothchild, since the fixing first began (up to NM Rothchild's withdrawal from the gold market in 2004, that is). Representatives of all five participating gold trading companies used to assemble on the premises of the NM Rothchild bank for each fixing.

The chairman suggests the first price at which supply and demand is queried by all fixing members. Each trading house checks and announces whether it is a buyer or seller at this price. If there are buyers and sellers at this price, each announces the number of disposable and ordered ounces. Should there be, for example, 100,000 ounces on the seller's side and only 20,000 ounces on the buyer's side, the price to be determined by the fixing is adjusted correspondingly. This continues until equilibrium is achieved between the buying and selling sides. If no balance can be achieved, the price can still be fixed, as long as the difference does not exceed the so-called range of discretion. This discretion volume is 10,000 ounces and means that any ounces "left over" will be split among the fixing's participants.

If you were lucky enough to enjoy the opportunity of observing a traditional fixing, you may have frequently come across the expression flag. All members of the fixing had British flags, union jacks, standing on their tables. As long as any of the flags were still standing, no price could be fixed. As soon as a trading house had announced a volume at a certain price, the representative would lay the flag flat on the table. But if a trading house ever changed its decision, for instance, by switching from the buyer's side to the seller's side, the respective bank's representative would exclaim, "flag," and stand the flag upright again.

Although there is no spread as such, buyers do pay a surcharge of 20 to 50 cents per ounce within the fixing. The seller pays no surcharge but instead receives a premium of usually up to 15 cents per ounce.

You have now been introduced to the trade options in the spot market. In contrast to stock markets, where an electronic posting is processed via Clearstream, for instance, the gold market deals with physical goods. As soon as a realizable material is circulated and if it is also simultaneously highly sensitive material, control and monitoring processes become unavoidable. The assay is a very work intensive and therefore also expensive process, which is one of the reasons why trust plays such a large role between the trade parties. Four things need to be assayed:

1. Accuracy of fineness hallmark
2. Plain material
3. Core test for alien material
4. Embossed weight

One of the methods employed to assess fineness is the black touch-stone process. The touchstones (a specific type of mineral) are rubbed against the material to dislodge minute amounts of gold. This produces a discoloration indicating the carat number or fineness, respectively. The process is obviously only a preliminary test to check for significant deviations. To determine the fineness down to within three digits after the decimal, samples have to be chemically examined in a laboratory. Amateurs have been known to buy all kinds of things in the past. Some counterfeits were discovered to be bars of lead or some other material alloyed with a thin gold layer. These kinds of fakes are of course easily detected by professionals. Sample drillings and a simple weight measurement would be enough to uncover the deceit. Much more professional counterfeits are gold bars with a wolfram core, for instance. Wolfram is the only metal whose density approximates that of gold and can therefore not be detected by weight measurements alone. Only a drilled sample will help in this case.

Trust is of the highest priority in the physical gold trade. Even a refinery's hallmark is easily copied. It is hence always a cause for suspicion should gold be offered below its market value. Gold always costs the gold price—and sometimes more—as the production of gold bars does not come for free. So-called molding costs will have to be charged as soon as the gold is cast into specific bar shapes.

The primary molding cost is usually covered by the gold mine itself, as it receives the price rate for molded gold from respective buyers, usually a bullion bank. Additional costs are incurred in the retail client trade where demand is reflected in kilogram or ounce bars. By the time a bar reaches its final shape it may have incurred multiple costs occasioned by several such molding processes. As there is generally no mass production, clients may be surcharged up to 80 Euro per kilogram bar.

There are three major gold refineries in Germany that primarily supply the jewelry and electronics industries, in addition to delivering to private clients. Pforzheim is the home of the Allgemeine Gold und Silber Scheideanstalt, now a subsidiary of Unicore in Hanau, which in turn is a subsidiary of the Belgian Unicore N.V. There is the long-standing Heraeus, which is situated in Hanau and the Norddeutsche Affinerie in Hamburg. Degussa, the Deutsche Gold- und Silberscheideanstalt, once the largest of Germany's refineries, is now trading under the name of Umicore, but still produces bars bearing the Degussa hallmark.

There is also a third physical gold shape besides the admittedly far more important bars and granulate: bullion coins. (The term "bullion coins" is only used for gold investment coins, not to be confused with numismatic gold coins.) Bullion coins are usually available at the current gold price with no or low surcharges. The coins are also traded without any numismatic collector's surcharge. One can still expect an extra charge of between one and three percent. In some cases, bullion coins can be traded at a lower price than their bar equivalent of equal weight. Faced with the choice between gold bars and gold coins, coins will turn out to be the cheaper as well as more customary variant in most cases.

One of the most famous gold investment coins is the Krugerrand. The Krugerrand is produced by the South African Rand Refinery. It has a particularly reddish sheen because it is not made from .9999 gold but also contains a tiny amount of copper. A Krugerrand weighs in at 33.8753 grams at a fineness of .9000. Each Krugerrand nonetheless contains one ounce of fine gold.

In comparison with the Austrian Philharmoniker, the U.S. American Eagle, the Australian Nugget and the Canadian Maple Leaf, the Krugerrand will usually turn out to be the least expensive. This is logical for several reasons. South African Krugerrand production is part of a gold export policy. Only native South African gold is used for coin production. Austria, for example, needs to first import gold to be able to process it. The Australian Perth Mint acts according to similar export policies as South Africa's Pretoria Mint.

These best-known gold investment coins have the obvious advantage of international acceptability and of frequently being traded VAT-exempt (since 1993, there is no VAT payable on gold bars and coins in Germany). The coin's name functions in a similar manner to a refinery's hallmark: It guarantees a reputation. In the interbank trade, Krugerrand coins are being traded with low surcharges, similar to gold spots. But their spread is greater and can sometimes reach several dollars.

In some countries, but particularly on the Indian market, the term gold bullion is also used for gold jewelry. Gold is frequently used as a currency or old-age provision in nations where the precious metal is part of important traditions or where it is regarded as a signal of wealth. In many Indian regions, the entire daily trade is based on gold, and goods, for instance, are routinely swapped for gold jewelry. Gold jewelry is not that suitable as an investment product, alas, as questions of fineness and intrinsic value also play a role besides acceptance factors. It is for this reason that gold jewelry should only be used to serve beauty (with the exception of countries traditionally tied to gold) and will not be treated as a cash product here.

THE MARKET FOR DELIVERY

Without the cash market, there wouldn't be a market for delivery. All forward transactions are based on the current spot market. But this raises the question: Why should there be a futures market for gold at all?

Futures markets for raw materials have existed for centuries, and futures products are still a customary instrument in the finance industry. The first institutionalized exchange for gold futures was opened in 1972. But even before that, futures products had been traded off-market for a long time, for instance, in the shape of forwards. Futures are—or at least that is the basic idea—an enterprise's instruments for hedging risks. This means institutions and persons who are liable to experience losses due to market price fluctuations have always had an interest in risk coverage instruments and forward products. Gold mine managements, for example, need to know what price they will receive for their not-yet-mined gold ounces in the following months and years in order to calculate the operating business in relation to running cost. It's not only a boon in successful company management if a company knows beforehand that it will be able to sell the mined ore at a pre-determined price that is already known today, but also limits the risk of potential price drops. And on the buyer's side, for instance in the jewelry industry, the parallel risk concerns unexpected price rises that would increase production costs. The jewelry industry therefore has the opportunity of hedging potential price hikes by using forward transactions. Hedging in this context means operative risk management.

THE FORWARD MARKET

The basic product, also for the futures market, is the forward market which operates on the idea that a supplier may sell a certain volume of gold at a pre-determined price and on a pre-determined date. The same applies to the buyer on the opposite side. Such a contract principally specifies the price the gold is to be bought/sold for at a future date and the contango (an interest premium resulting from financing cost, USD interest rates and the gold lending rate, depending on the volume and maturity. See below). Let me first explain the exact function and modification of a forward, which is also simply called a swap. A gold market swap can have the following meanings:

1. The simultaneous purchase of a cash position and sale of a forward (or vice versa).
2. A switch of gold positions (Loco swap).
3. An exchange of physical material for the same material in other sizes and qualities (size/quality swap).
4. An agreement about swapping a variable gold price for one that is fixed and agreed upon.

A basic distinction is being made between outright forwards and swaps. An outright forward exclusively involves the future sale or purchase of a position. Whereas a swap takes the equivalent opposite side to a futures position in cash, i.e. a forward purchase is balanced by the sale in cash. The function of a classic outright forward is described in the example below:

A jewelry manufacturer planning to produce jewelry with a volume of 10,000 ounces (large manufacturers employing numerous goldsmiths can easily cope with this amount) in three months' time can predetermine the price that will be valid in three months' time by buying a three month forward. The trader will quote a price, which is expressed as a percentage, of let's say 1.02 to 1.12 (with an assumed gold selling rate of $420.20 USD). The calculation is actually quite simple: Some of the figures represent the bid-offer spread and the others represent percentages relating to the gold price. Based on a chronology of 30/360 days the following three month premium results: $420.20 USD * 0.0112 * 90/360 = $1,177 USD

As the jewelry manufacturer would like to buy the gold, the trader quotes the asked price, i.e., the 1.12 percent. The result is a three-month gold price of $421.37 USD. The difference is called carry premium. The important question now is how exactly this percentage quotation of 1.02 to 1.12 is arrived at and how banks or traders safeguard themselves when quoting such a price.

The percentage quote is called the forward rate and is principally arrived at in the following manner: Forward Rate = USD interest rate* – gold lending rate.

In the example above, the trader is committed to selling the gold to the jewelry manufacturer in three months' time at a price of $421.37 USD. This means that the trader has to safeguard the transaction by buying the gold on the day of the trade at the underlying price and storing it for three months. This is where the trader incurs his costs: He first needs to spend the $420.20 USD per ounce in order to purchase the gold. He therefore finances this expenditure in the inter-bank market at the USD interest rate. This interest is initially due as cost. As he will only need the gold in three months' time to honor his agreement, he can lend it out in the inter-bank market in the meantime. He hence also acts as a gold lender in the market. The lease rates are markets in themselves, which are based on parameters other than the USD interest rate. It mainly depends on the market supply of a material and insurance costs. To

* The reference interest rate LIBOR is most commonly used

put it more simply, the more gold there is available for lending in the market, the lower the lending cost or yield, respectively, will be. This is why the trader must know the lease rate applying to his three-month-gold before he closes the forward deal. As the lease rates usually range below the USD interest rate, the formula has a positive result. This positive result is called a future price surcharge or contango, i.e. a premium. It comprises a rising scale of futures prices, a positive futures curve, which means that longer-running forward contracts are more expensive than short term ones. But in some cases, the surcharge can also turn into a discount. This happens when the lease rate exceeds the USD interest rate.

Such an event will occur when the market experiences a gold shortage. Bottlenecks are created by excessive demand or short term supply lapses. In these cases, forward prices would be lower than the current market price, the spot market. This is called a backwardation market, where a discount applies.
The scenario described above equally applies to the sale of forward prices.

Forward trades have some peculiar characteristics that need consideration. As their interest calculation is based on two markets, potential interest rate differences need to be accounted for. The forward price calculation is usually based on actual/360. This means that a month will be calculated as 30 days. Other markets base their calculations on 365 days or various other parameters.

One frequently comes across the term "forward leg." As the swap comprises two components—a cash and a forward transaction—the term forward leg is used to describe the forward component. On the opposite side, the forward leg is balanced by the physical part of the transaction, the cash component.

The term "cash-settled" swaps denote swaps that are not based on a physical delivery, but on an evening up. Here, the forward price difference will not be settled in physical goods, but financially, at the current price.

Forwards are principally traded on a monthly or annual basis. But they can basically be traded with any term of maturity: from a few days up to several years. Short-dated forward contracts can have a lifetime of just a few days. Broken date forwards are futures contracts that fall outside the typical weekly or monthly rhythm. This flexibility of forwards ensures that virtually any demand can be traded and hedged in due time.

In the absence of other agreements, a forward contract's value date is set at two days (bank working days in London or New York, respectively) after the deal is closed, just like with cash trades. A further peculiarity is the spot-deferred contract classically used in mine hedging. Here the typical forward is tied to an option. The term deferred means that the delivery date has not been defined. The basic contract resembles a forward, except that no delivery date is set. The seller is granted the option to defer the due date by rolling over the contract. This is done on the contract's due date. Not only can possible losses or profits be delayed and realized at a later date in this way, but mine production volumes can also be sold beforehand while their delivery is delayed until they are actually mined.

The forward market is, like most gold products traded, an OTC trade instrument, which means that deals are closed bilaterally and do not normally appear in any statistics. Due to the direct nature of the business, the trades can consist of specialized and amended contracts. They are not standardized.

The futures market is, in contrast, characterized by standardization. Only fixed terms are traded. It can be very important for traders to monitor the relationship between futures and forwards in order to profit from possible price differences. The resulting trade would be an arbitrage. We will return to this in greater detail in chapter 5.

THE LENDING MARKET AS A COMPONENT OF THE SETTLEMENT PRICE

As we have seen in the previous chapter, the lending market is a fundamental part of the futures market. The lending market puts market participants in a position to lend or borrow their gold in cases of shortage at interest cost. Gold positions that are not required by a trader, a bank or an asset manager for a certain period of time can thus be passed on to interested third parties via the market. A classic example: a bank issues certificates that are secured by gold. Empirical values indicate how much of this gold is usually left as a residuum, i.e., is not called in by clients prematurely before the term expires. The bank can put this gold to work by offering part of this residuum to the lending market.

The party furnishing the asset to the market is called the lender and the party who requires the gold on loan is called the borrower. One can principally say that the lending rate curve tracks the supply and demand situation. This means that a higher lending rate is to be expected if there is a supply deficit, whereas lending rates can even turn negative in cases of excessive supply. Negative lending rates are obviously caused by insurance and storage costs payable to the borrower by the lender. But this extreme scenario only occurs if there is an enormous oversupply of gold. Since the summer of 2004, lending rates have been negative at the shorter end of maturity terms, which means that this extreme scenario is currently being experienced by the market.

Lending rate calculations can be "un-calculated" using the forward rate, because

USD interest rate - forward premium = gold lending rate

(Please note: The bid-offer spread always needs to be included in this calculation.)

Lending rates can principally never be represented lineally on a time curve. Bottlenecks, should they occur, often come about in the first few months and are therefore frequently short term. But longer maturity terms are usually more expensive because each lending transaction also represents a credit risk. The following illustration shows the volatility of lending rates.

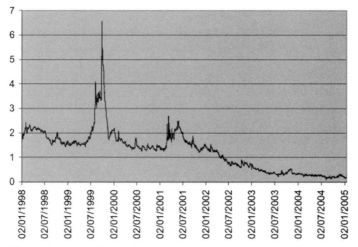

Ill.: Lending Rate 12 Month Maturity
Source: *Bloomberg*

Lending in this case means granting a credit because the gold
physically changes hands while its direct or indirect owner takes
the part of the borrower. This is also the reason why there are two
distinct lending rates for the inter-bank trade and the industrial
client trade. As most jewelry manufacturers' credit-standing is
below that of a bank, their lending rates are raised by an additional
solvency risk, accordingly. By its very nature, the lending trade
is the domain of central banks because they happen to have large
volumes of gold at their command that can not often be disposed of
but may instead be lent to other market participants.

A distinction is sometimes made between gold lending and inventory loans. This is, in fact, a combination of lending and forward
that goes by another name. Inventory loans are normally granted
to the jewelry industry. It means that a bank extends credit to a
jewelry manufacturer by supplying physical gold. The manufacturer can now produce the jewelry and will repay the loan on the
day he sells it by buying the gold back in the market. This way, the
manufacturer can avoid a price fluctuation risk, while simultaneously financing his production.

It should finally be noted that lending interest can only be charged
and paid for unallocated gold. Only if the material is stored collectively can it be leased and earn interest. Allocated storage excludes

this application. In addition, the lending market predominantly trades in Loco London, meaning that other marketplaces are subject to different price scenarios.

FRAs (Forward Rate Agreements) and IRSs (Interest Rate Swaps) are financial products that have entered the gold market from the general interest-rate and financial markets. They are used to control the risks and fluctuations of gold lending rates. But as these products have been on the wane for several years, they shall not be explained here in detail.

THE FUTURES MARKET

A futures market is a standardized forward market. Futures markets are also rooted in hedging. Futures augment and substitute the OTC forward market. Augmentation in this context means that certain minimum standards and prerequisites apply to the trade in a futures exchange so that relatively high standards can be ensured. And substitution in this case means that the products can be traded more easily and transparently.

General futures exchanges have been around for over 100 years. But the very first gold futures were only traded in 1972, in Winnipeg, Canada. Although gold futures are no longer being traded in Canada, they did manage to establish themselves globally, particularly in the U.S.

A futures contract is a standardized agreement to buy or sell a certain volume of a certain material of a defined quality during or at the end of a fixed period of time. This definition is very similar to that of a forward—except for the specification of standardization. These contracts are usually available with various maturities expressed in months. In New York's Commodity Exchange (COMEX), front-end, or nearby futures contracts, respectively, are most actively traded. Whereas in the Tokyo Commodity Exchange (TOCOM), the most actively traded gold futures are back-end futures with a maturity of approximately one year.

COMEX's market mechanism has not been changed in decades and instead refashioned the gold market in its own shape. In the wake of the computerization of trading, numerous attendance exchanges have been replaced by electronic trading platforms. But the COMEX has successfully resisted this trend and the price fixing is still performed on the floor by open outcry trading. Deals are closed as soon as a bid and an ask price match (i.e. as soon as a buyer and a seller agree on a price). The brokers on the floor usually maintain direct contact with their clients who are mostly banks or funds. A bank's trader communicates orders directly to the broker so that they reach the floor within seconds. A distinction is made between market orders—orders to be executed in the market at the lowest (in case of a purchase) or best (in case of a sale) market rate—and so-called limit orders, which are only addressed at a certain price level. The bank's trader receives a confirmation by telephone immediately after the deal is closed. Because this type of trading heavily depends on verbal and physical signals, it is quite prone to error—or at least more so than trades entered into via electronic platforms. This is why all telephone communications are recorded and orders are given a time stamp, which ensures that a process can be re-construed after the fact. The floor is also under constant video surveillance so that mismatches resulting from hand signals can be checked later.

Most brokers act on the floor independently, but some represent larger banks. In most cases, brokers act at the behest of a client. But American brokers are also permitted to hold their own positions, as opposed to the regulatory situation in Germany. If they act on their own behalf, they appear as locals. Typical signs are small trade volumes and brief holding periods. For instance: If a broker sees a larger order entering the market, he can pre-empt the large buyer with small quantities in order to partake of the price rise resulting from the client's order volume.

As I have already mentioned, futures markets were first established to supply producers and other market participants with hedging instruments. This is also the reason why only a tiny portion of all contracts traded actually result in a physical delivery, which would mean that the seller is forced to provide physical gold. Ninety-eight percent of all contracts are usually executed on a purely financial basis as margin business. As a result, the market attracts a large number of speculators who usually lack gold holdings and are not able to absorb any.

GOLD

Characteristic features of forward markets like backwardation are also to be found in the futures market. Sometimes even in a much simpler form because these exchanges offer contracts with very long maturity terms. But in addition to these common features, futures exchanges also comprise some unique characteristics. One of these is the depots, for instance. Futures traders holding selling contracts are, by definition, required to deliver the gold on the day of maturity, unless the contract was closed or a bilateral settlement of the deal could be achieved beforehand. Commodity markets—and this distinguishes them from financial futures markets—nonetheless still maintain a close connection to the underlying material. These markets are controlled by supply and demand. Delivery bottlenecks and situations of excessive supply might occur and their interaction creates the price. To maintain the connection between the futures market and its goods (i.e. the gold) and therefore create a situation where, at least theoretically, all contracts could be physically fulfilled, the COMEX established warehouses that are regarded as official places of delivery. In these depots, various traders hold physical gold positions which could be used for deliveries, should the case ever arise. The inventory of these depots is an important indicator for an exact analysis of the market situation attempting to determine whether there is a shortage or an oversupply.

The open interest principally augments this information. It is defined by all open contracts in the market; that is, contracts still outstanding or awaiting execution. The open interest figure indicates the number of market participants currently holding positions.

The Commodity Futures Trading Commission (CFTC), which regulates the futures exchange COMEX, among other duties, splits all active market traders into two groups: non-commercials and commercials. Non-commercials are market entrants who are not trading the underlying value gold—speculators or long term investors, for instance—while the term "commercials" denotes hedgers, producers and consumers of the raw material. The distinction emphasizes the fact that commercials and non-commercials enter the market with different intentions. Commercials are primarily interested in risk control, which they partly try to achieve by establishing long term positions, whereas the behavior of speculators serves short term goals only with positions that are frequently just held for a few weeks. The CFTC publishes the respective positions of both groups weekly (on Friday afternoon). This puts outsiders in a position to gage the behavior of speculators in relation to that of mines and producers. The operative figure is that of the net commercials. This number shows the sum of all positions held by both groups. If there are many net long non-commercial positions, one can safely assume that a large number of speculators expect the market price to rise. The following illustration indicates the high volatility of these positions. One can also clearly see how market movements track the positions of speculators.

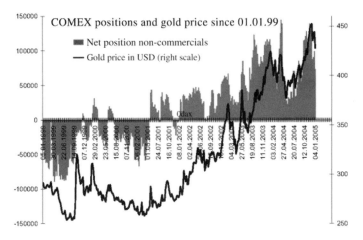

Ill.: COMEX CFTC Gold Positions
Source: *CFTC*

The calculation of futures prices is the same as for forward prices. But there are differences in the individual cash flow of the two contract types. With a forward, the total amount usually only flows from the buyer to the seller on the day of maturity. The seller carries the contrahent risk in the meantime, during the period from entering the forward trade until it is due. The seller needs to control this risk separately. The futures side works with margins (payments deposited as a security for the contract entered into). There are three types of margin: initial, variation and maintenance margins.

To be able to buy a contract, one first needs to pay the initial margin to the exchange. The payment is intended to cover a certain percentage of the total contract value. Margin payments usually only represent a very small fraction of the overall contract value, though. Depending on market fluctuations, each contract entered into will either increase in value or incur losses. This means, depending on the direction of the market development, the exchange may require a supplementary payment—an addition to the amount already paid in and deposited. The exchange will in this case perform a margin call in order to collect this additional payment, the so-called variation margin, from the contract owner. Whereas a maintenance margin will only be called if a gold price or certain price level determined before the contract was made has been reached. The exchange also prescribes specific minimum margin levels to bankers and traders active in the market. But trading houses are entitled to charge their clients different call payments.

The following example shows a typical futures transaction. The margin payments conform to the current conditions at COMEX in New York.

The gold futures price for the May 2005 contract is $408.50 USD (median price). The contract is for 100 ounces. The total value of the long position is therefore $408,500 USD. The COMEX demands an initial margin of $2,500 USD for this (i.e. 0.6 percent of the total value). If the gold price now drops by $8 USD, the position's buyer stands to lose $8 USD * 100 ounces = $800 USD. This means that the trader's initial margin has been reduced by $800 USD. As the margin level would have been crossed even at only $200 USD below the initial margin, hence causing a margin call, the buyer now has to pay a variation margin amounting to $500 USD. If the futures price had climbed, the buyer would of course not have incurred any loss. As a matter of fact, his position would have increased in value.

The example shows that a relatively small amount of money is able to leverage quite a sizeable volume of gold. The payment demands via margins are designed to ensure that the buyer/seller would in a worst-case scenario be able to fulfill his financial obligations.

THE EFP MARKET

EFP stands for Exchange for Physical. This instrument (which is actually nothing more than the forward contract's interest calculation) was created to optimize trading possibilities between futures and cash markets. With EFP, a long position can be switched to a physical position and vice versa without selling the contract or waiting for it to mature. In this procedure, the client is not subject to any outright risk, i.e. there are no uncovered positions.

The following example shall illustrate this. A jewelry producer is hedging via the purchase of a futures contract. As the manufacturer originally expected that he would require the gold in 10 months' time, he bought a contract with this maturity. But now he discovers that he already requires the material after only five months. If he were now to cancel the position and thus be forced to buy the physical gold in the jack, he would have to pay for the spread twice over, increasing his expenditure accordingly, and would also incur a higher operational risk. The manufacturer hence contacts his bank and requests a two-way price for an EFP. The bank would then quote a price of, let's say, $0.55 to $0.85 USD. The producer can now switch his futures position for a physical position in a single transaction for the price of $0.55 USD (the amount of $0.55 does not represent the direct cost, but only the interest expenditure he retrieves prematurely). Another very important advantage is that the manufacturer can perform this procedure regardless of the exchange's operating times.

What underlies the EFP price quote is ultimately just the difference between the spot price and futures price cost. But the larger banks maintain separate EFP account books, i.e., cash as well as futures positions, which they can efficiently control internally. The EFP market is also useful for quickly putting larger positions that have built up in the bursary (physically) on a more transparent and liquid futures exchange. The product EFP therefore mainly augments the trading options of arbitragers and inter-market traders. But it is generally important to have a detailed knowledge of exchange regulations, even if this process occurs beyond the pits, as it is still regarded as an exchange trading transaction.

THE DERIVATIVE MARKET

The derivative market is a relatively recent development. Gold options (i.e. the classical form of a derivative) have only been traded for approximately 30 years.

Derivatives are products whose price development depends on other products. They are hence derived from price developments via various parameters. Options are not only the best-known derivative, but also the gold market derivative that is most actively traded. The options explained below are typical vanilla options; that is, they are not exotic and complex constructs like digital or Asian options, which often require mathematical operations of a highly complex nature.

OPTIONS

Although forward and futures products can be, by their nature, very efficient hedging instruments in view of unexpected price fluctuations, the product range needs to be rounded off by the world of options.

Options are derivatives, meaning their price development depends on a base value. Gold options are principally the same as other financial options. In addition to the OTC options, there are also listed ones, for instance, at New York's futures exchange COMEX (these options are based on an underlying futures contract and not on the spot price, as is customarily the case with OTC options).

A basic distinction is made between buyer's options (calls) and seller's options (puts). The acquisition of a call entitles the buyer to purchase a previously specified volume of gold at a previously specified price on a previously specified future date. The buyer always has an entitlement, but is not obliged to make the purchase. The acquisition of a put entitles the buyer to sell a previously spec-ified volume of gold at a previously specified price on a previously specified future date. In this case the buyer always has a right, but is not obliged to sell.

The options buyer obviously has to pay something for the acquisi-tion of these entitlements. This price is the so-called premium. The option premium is often expressed as a percentage of the underly-ing value or as a volatility percentage. The premium's calculation is based on six parameters that are included in various option price models. The most famous of these methods is the one developed by Black/Scholes in 1973, which includes the following parameters:

1. Current price of the base value, the gold spot price
2. Exercise price, strike price
3. Term of the option

4. Interest rate relevant for financing gold
5. Interest rate of the underlying value, in this case the gold lending rate
6. Implicit volatility

Except for the volatility, all of these parameters are objective, i.e., they are openly obtainable in the market. The data need only be integrated in the Black/Scholes equation. Only the volatility parameter requires special attention, as it is mainly the volatility that will determine the value of an option and its development.

What is volatility, and how can there be two different volatilities? Volatility is a ratio which is expressed as a percentage. Historic volatility reflects a base values' price movements within a certain period of time. It can most easily be compared to standard deviation. But options prices generally include the implicit volatility. This is not based on historical values but rather on an experiential value, i.e., it expresses the price movements that are expected in future. In this sense, it is also a subjective evaluation of each individual market participant. This also explains the price differences occasionally occurring between options from different banks.

The options jargon abounds in terms "like in-the-money," "at-the-money," and "out-of-the-money." These are used to denominate the value of the options. An option is hence in-the-money if it has a value at the time of being exercised (i.e. if it makes a profit). An option that's at-the-money is approximately at the break-even point. And an out-of-the-money option is worthless when exercised, as it does not have any inherent value. This means that the underlying material (the gold) can be bought and sold at a lower price in the market than the pre-defined strike price of the option determines. Let me show you some examples for options.

The most frequently occurring type of option is the plain vanilla call. Semiconductor manufacturers who require gold for their bank contacts, are very price sensitive due to the intense competition. They know their exact production volumes, but from time to time they place orders that are of a unique nature. They need to hedge against rising gold prices in order to be able to continuously offer their products at a stable price. Buying a forward would be unwise in this situation, as the manufacturer does not yet know if and when the deal will be closed. And if they were to buy futures or even the material itself they would run a permanent price risk. But with a buying option they can safeguard themselves over a pre-defined period of time. Initially, only the premium will have to be paid and not the underlying material's total value. If the option is abandoned, the premium will create some expenditure, but there would be no obligation to purchase the gold. The purchase of selling options is typically performed by gold mines. A mining company knows its production cost per ounce approximately. The sale of each ounce therefore needs to at least cover the production cost without incurring a loss. If a gold mine should, in a worst case scenario, wish to realize this cost item as earnings while simultaneously participating in price rises, it will purchase a put, enabling the buyer to sell the gold at production cost in the worst case scenario. In this case, the expenditure is again limited to the cost of the option's premium. And once again the purchase of puts does not create any obligations.

If one can buy options, one can also sell them, thereby granting an entitlement to third parties while making a commitment oneself. Whoever sells a call or a put always enters into an obligation and is always at the option buyer's beck and call because it is the buyer who decides whether to exercise the option or not.

A further distinction is made between American type and European type options. European options may only be exercised at the end of their term, whereas American options can also be exercised before they reach maturity.

The options traded in the circles of the London Bullion Association are usually quoted in absolute amounts or in volatility percentages. Just like in the spot trade, the availability or exercise date is in two days' time. Principally, either a physical performance or a cash settlement is possible. A physical performance would need to be in Loco London drawn from unallocated accounts.

OTC options principally harbor a counterparty risk, i.e., the risk of a contrahent's failure. The risk is usually run by the option buyers, as they are the ones who'd be entitled to receive something should an option be exercised. The risk in this case concerning the ability of the other contract party to fulfill this obligation.

Apart from the plain vanilla variety, there are further options and options structures which can be a lot more complex. I shall just mention two examples here: Asian and barrier options. Asian options are not based on a fixed strike price but on a price average during a specified time period. Barrier options are based on price ranges and their value is determined by whether they reach these benchmark values or not.

The price calculation groundwork applying to OTC options can of course be used in a similar manner to calculate exchange traded options. The latter are available at COMEX in New York, for instance. But the options at these exchanges frequently refer to future contracts traded at the same exchange. In this case, the OTC options would refer to the gold spot price, whereas the options contracts traded at COMEX would refer to the future. Futures prices and spot prices are obviously quite similar in their price development, but the strike prices differ due to the interest expenditure (cost of carry) included in the price of the futures contract. There are several other standardized terms relating to contract term, settlement and the exercising of options, as OTC options can be fully adapted to a client's individual requirements. It is this customizing that also facilitates the management of larger positions. A large mines' yearly output cannot simply be traded via exchange options without anyone noticing, for instance. In such cases, OTC options offer the advantage of discretion, while simultaneously promoting market transparency.

This general benefit emerges more clearly if one considers the aforementioned flexibility of these instruments. They can be used as a hedge for large, but also for small positions. The liquidity of the markets also permits a pure trading strategy. Entering and closing positions in the markets is usually possible without running any operational risks.

RETAIL MARKET PRODUCTS - ULTIMATE CLIENT TRADE

In the past, private individuals and low volume clients only had limited access to the gold market. Although the products described above had already been established in the markets, they were not accessible to this clientele. Even as recently as 20 years ago, the simplest method was to buy gold coins or small gold bars, in spite of the higher taxes and fees.

During the last five years, derivatives were the financial world's dominant product. Futures exchanges and warrant exchanges like the European warrant exchange Euwax in Stuttgart, Germany, have only been able to enjoy their turnover increases quite recently. Even this segment, which has been exclusively reserved for institutional investors for a long time, now grants small traders access.

Today's investment products are more varied and diversified than they have ever been before. Let me introduce some of the product categories that appear to be customary nowadays in terms of turnover.

CERTIFICATES

Certificates are derivatives, just like options, and embody entitlements or calls. These calls or entitlements are not firmly defined and can therefore appear in all kinds of forms. Futures, physical claims or even option structures may be certified (i.e. embodied) in this way, to mention just a few.

The structure of the very first certificates was quite simple. They were exclusively used to facilitate the trade in certain underlying materials in smaller units. They are also known as price trackers, participation certificates or tracers. Parallel to options the term "plain-vanilla-certificates" wouldn't be out of place here. Their mechanism can be explained quite quickly.

If you take, for instance, the price of gold as it is quoted in the international inter-bank market per ounce in USD, a gold participation certificate would represent one-tenth of the gold price denominated in a national currency. The investor hence has the opportunity to participate in the gold price for a pretty reasonable outlay—one-tenth of the value of an ounce of gold—without requiring separate currency transactions. That is the call, whereas the issuer's obligation is to keep track of the price development and credit the certificate buyer accordingly. Buyers are not usually entitled to a physical delivery. Potential investors should also be aware of the fact that they buy a borrower's risk, as the entitlement is based on an issuer's promise which means that the risk hinges on the standing of the respective issuing house.

A relatively new transmutation of this basic form of certificate is so-called Quanto-products. Quanto-products are certificates that include currency option structures which revalidate the currency risk of investments. Because precious metals are usually traded in USD, a Euro-investor runs a certain currency risk. These products are therefore particularly useful for investors who do not invest American dollars. Quanto certificates are more expensive than plain vanilla certificates, because the currency hedge via options creates additional expenditures.

The basic idea behind the development of these certificates is the following: a market that is usually closed to certain investors due to market restrictions, geographical issues (time differences) or simply because of the high cost associated with it, is to be made more accessible. Certificate constructs can also serve as all-inclusive solutions, as we have seen in the case of the Quanto-products.

By now, even futures products may be vested as certificates. These products can be tricky and are often regarded as financial innovations because their construction needs to ensure that investors don't reap negative performances, but may in a worst-case scenario lose their entire investment capital instead. Also customary, if in a smaller number, are products that are tied to Asian options. These products are usually emitted at leverage on the underlying, while the leverage and participation calculations are based on a key date rather than on a continual daily pricing.

It should be mentioned in closing that the certificate market currently offers a large variety of structures and products. Their number and complexity is constantly increasing. One of the results is that the circle of private investors is now able to access a significantly larger amount of information on the gold markets and their products. But the true extent of demand will, on the other hand, never be apparent from the gold market itself, as a direct intervention in the physical market is not always required.

WARRANTS

Warrants can most easily be compared to classic OTC and exchange options. They principally differ in the possibility of not exercising the optional right and in that they are a pure buyer's market (it is not possible to "sell" warrants and their target market is different to that of options).

In Germany, warrants are traded at warrant exchanges and OTC. In this respect, Euwax is currently the market leader for this type of derivative product. Purchases in the OTC market can also be transacted directly with the contracting party.

The function of a warrant is similar to that of an option. But they do offer an advantage over OTC options as the products are traded in exchanges, which means that buying and selling prices are permanently provided. And they facilitate the trade in smaller volumes. As warrants are standardized it might not always be possible to find a product that exactly meets the requirements.

METAL ACCOUNTS

In Germany, this product category has been around for decades and is still part of the product range of most banks. There are two different types of metal account: metal account with allocated storage and metal account with unallocated storage.

Metal accounts are basically comparable to currency accounts, i.e., the client opens an account at his bank that either carries the name XAU for gold or XAG for silver. The client then orders the desired volume of gold in ounces or grams. If required, clients may also have their metal account denominated in a specific currency. The credit institute subsequently buys the gold on behalf of the client and electronically posts it to the customer account.

Up to this point, there is no difference between allocated and unallocated storage in metal accounts. If the metal account is for unallocated storage, the institute buys the gold intended for their client in the bank's name and on the bank's account. The gold is thus directly owned by the bank and also stored by it, whereas the client who bought the gold for his metal account with unallocated storage only has an entitlement against the bank (i.e. a claim to have the gold delivered). In cases of insolvency, this kind of structure results in the client's claim being subsumed in the total of distributable assets, which means that the client runs the risk of the bank defaulting. If worse comes to worst, the deposit is not secured by the deposit protection fund and the capital is completely unprotected in case of insolvency.

Allocated storage eliminates this default risk because the purchased metal is taken out of unallocated storage to be separately stored in the client's name. This structure is obviously safer, but also much dearer, as the separation of bars according to a new numeric assignment is performed at the client's expense.

Metal accounts have several cost-related advantages. In the past, gold price speculators could quite reasonably buy and sell metals deposited in accounts electronically, thereby avoiding the cost associated with a physical transaction. In Germany, metal accounts were particularly propitious in tax terms, as VAT was still due on gold, and as long as the metal was not delivered physically, the investment was VAT exempt. Metal accounts still make great sense for silver, as ownership of physical silver is subject to VAT. Silver, for instance, can also be traded fiscally neutral in metal accounts. Tax demands only occur in cases of physical delivery.

GOLD LOANS

Gold loans are predominantly structured products. Their buyers can either participate in the gold price or they receive interest at a constant rate. This interest may result from the lending rate.

The issuer of gold loans tied to the lease rate buys the gold for the client product and puts it out on loan in the market at the current lending rate. In this case, the lending rate acts as a coupon on the paid-in capital. This is a gold loan in the classic sense, but its construction is to be viewed very critically in terms of risk, as it creates a contrahent default risk on the ultimate borrower's side, in addition to the gold loan's inherent issuer risk. Furthermore, this kind of product only really makes sense as long as a high lease interest is available in the market. This is currently not the case.

Other variants involve the purchase of zero bonds and the purchase of options representing the participatory component. In these, the client's capital is invested in interest-producing bonds whose earnings will reach par value at the end of duration (zero bonds). The difference between the interest-producing bonds and par value is used to represent the option component. These relatively small amounts are invested in a call option. Because the amount is only sufficient for an option representing a lower ounce embodiment than the paid in capital would, this loan construct can only embody a partial participation. Such a product could have the following characteristics, for example: Issuer's rating AA, capital security 100 percent, three-year maturity, coupon calculated from 30 percent positive participation in the development of the gold spot price.

This type of gold loan is largely disconnected from any physical investment. Instead, a double issuer risk is embodied, i.e., that of the gold loan issuer and that of the zero bond issuer.

Although gold loans are rather conservative products, they can be useful in providing a classic bond portfolio with the required diversification and balance.

REVERSE CONVERTIBLES

Despite their name, the function is quite simple: The issuer sells a put with a particular strike. He can use the premium thus received to offer the buyer of a reverse convertible interest payments. These are balanced by the risk that the gold price may be lower than the strike level on maturity and the buyer will hence receive an equivalent number of gold ounces instead of a financial payment.

Let me describe this product, which is a stock-in-trade of asset management, in greater detail.

Product example: reverse convertible, annual interest 12 percent, price basis $410 USD.

A client invests $1,000,000 USD, which, at a price of $410 USD per ounce, represents 2,439 ounces of gold. The issuer sells a put for 2,439 ounces with a strike price of $410 USD. For this, he receives a premium of $120,000 USD, which he can pay out to the client in the shape of the interest agreed upon.

Scenario 1:
The gold price climbs to $460 USD. The sold put expires valueless (i.e. the sale of the put does not result in any obligation to exercise or any purchase commitments). The buyer receives his invested capital of $1,000,000 USD back and also earns the 12 percent interest from the sale of the put.

Scenario 2:

The gold price falls to $380 USD. The sold put is exercised by the counterparty and the buyer of the reverse convertible is obliged to buy 2,439 ounces at $410 USD per ounce, the equivalent of his invested capital of $1,000,000 USD. The interest remains the same.

This product is particularly well-suited for scenarios where sideway price movements are to be expected. The main demand comes from clients who require continuous interest payments.

Apart from reverse convertibles, there are also discount products. Here, clients buy long positions at a discount with a cap. The product is also a play on volatility, as both products sell options, they sell volatility. Or to put it simply: If volatility increases, the options become more expensive. The same applies to the steady interest as well as the discount.

Discount and reverse convertible products are good examples of the standardized volatility products currently customary in the markets. Several further variations of increasing complexity can be developed from these without altering their basic structure significantly.

The gold market is special in many ways, but its greatest distinction probably lies in the large variety of local peculiarities and idiosyncrasies of gold trading—and this despite its global, 24-hour nature. So one not only has to deal with the classic differences regarding the currencies of various gold trading regions, but also with varying trade practices concerning the metal's purity and units of measure.

THE INTERNATIONAL GOLD MARKETS

THE INTERNATIONAL GOLD MARKETS

THE INTERNATIONAL GOLD MARKETS

THE INTERNATIONAL GOLD MARKETS

THE INTERNATIONAL GOLD MARKETS

- II -

In India, for example, gold is traded in its purest form (99.99 percent pure), the unit of measure is Tola and the method of payment is the Indian Rupee. The Hong Kong Chinese prefer their gold less pure and measure it in the unit Tael. To pay for it, they use the Chinese Yuan. This tiny example serves as a perfect introduction to the following chapters, in which I shall try to describe local gold market conditions and specialties in much greater detail.

Each of the 20 gold markets below is described in exactly the same manner: first the historic background is sketched out, then the current market structure is thoroughly explored and finally the international importance is assessed.

GREAT BRITAIN

The United Kingdom is the most important gold market historically. Not only did gold fixing start here, but England is also the
home of the London Bullion Market Association which shaped the
physical gold trade with its professional regulations.

The very first bank activities go back to the year 1684. The first
bullion bank—Mocatta & Goldsmid—was even established before
the Bank of England. The London market's later significance was
also a result of the early introduction of the gold standard during
the 18th century. After the standard had been introduced, the Bank
of England became a permanent gold buyer in order to mint coins
and continually build up reserves. Wherever gold became available in the markets, the Bank of England stood waiting to bid for
it. This resulted in a growing number of gold sellers entering the
English market because they could be certain to sell their material
at a fair and reasonable price.

Next to the demand created by the Bank of England, the Commonwealth naturally provided an ideal base for trading gold extracted
from the colonies in England herself. Thus, the world's two largest
gold centers were initially tied to England: on the one hand South
Africa, at the time the largest gold producer and a colony of the
English crown, and on the other India, still one of the largest
gold buyers worldwide today. In British Colonial India, gold was
primarily used in the jewelry industry. Another member state of
the Commonwealth, Australia, also contributed to the movement
of gold to London. Australia had operated gold mines from pretty
early on and delivered the gold produced to the United Kingdom.

Due to the continuing gold influx, the gold standard could function in this manner without experiencing any greater difficulties.
Other European nations followed the example set by the UK and
introduced their own gold standards. But England played a leading
role right from the start as far as the gold trade and the distribution
of gold via central banks was concerned. As a result, many trade
transactions first went to England in order to be ultimately settled
in Europe.

Quite soon, the Bank of England acted as a gold depository for
other central banks, mostly of nations that were members of the

GOLD

Commonwealth then and partly still are. Large gold reserves would therefore be stored in the Bank of England in the name of third parties. And that is where the particular historic significance of the trade practice Loco London stems from. Over the years, the gold price traded in England became the international reference price. All gold traded around the globe was increasingly aligned to this English quote.

Because of colonial rule, South Africa had, in previous centuries, offered nearly all of the gold it produced on the English market, thereby creating a very close bilateral relationship. The trade management was performed by the bank NM Rothchild which initially held sovereignty over the acquisition of gold. In the beginning, all African gold was transported to London in the shape of Doré bars, i.e., in a preliminary stage of the end product to be then refined by Rothchild, Raphael and Johnson Matthey. The refineries frequently sold the gold to the Bank of England or offered it for sale to London gold traders. This relationship continued perfectly and the distribution and bank financing functioned smoothly until the year 1968. The crisis of 1968 introduced an imbalance that would result in a new gold market the likes of which had never been seen before.

Before the gold market we know today came into existence with its floating (free gold price), the price used to be fixed by the Bank of England. No individual, free price quote was possible. The traders of the time were pure intermediaries. Trades were either transacted directly via the Bank of England or via refineries who frequently granted credits to mines in order to secure their later purchase of the gold produced.

It was hence the Bank of England that formulated the very first gold trade regulations in the wake of the increasing trade activity in the precious metal. In the same period of the introduction of these standards, the time-honored bank house NM Rothchild also introduced the first standard bars. Up to this point in time, the Bank of England had traded and insisted on accepting 200 ounce bars (then still being bought and sold at a purity of .9160). The 400-ounce bar with a purity of .9950, which is still being designated as a standard bar by the London Bullion Market Association today, dates from this Rothchild period.

The gold trade of the 1920s was characterized by two important events: In 1915, the Bank of England abolished the gold standard and the bank house NM Rothchild initiated the first gold fixing.

The end of the gold standard and the outbreak of the First World War in 1914 paralyzed the trade. Gold produced overseas usually remained there, as transports were hazardous during the war and also because the demand dropped dramatically. Only after NM Rothchild started the fixing in 1919 did the gold market revive again. The Bank of England and the banker NM Rothchild collaborated closely to finance South African mine production and to market the acquired gold. At the time, Rothchild had once again managed to occupy center-stage in the international gold trade. The fixing was initially performed via telephone, but was later carried out during a formal meeting that convened every morning at 11:00 am on the Rothchild premises. Once the fixing was born it would persist in its original form until 2004.

There was one interruption during the Second World War, when no fixing occurred. But the English gold market could once again benefit from a financial policy decision. This time around it was the Bretton Woods conference in 1944. The agreement ratified there defined a new gold standard with the Bank of England acting as the clearing house. The creation of a new international currency structure with a gold price fixed at $35 USD was strongly influenced by the good experiences European Nations had made with the gold standard that existed until 1915. During the period from 1871 to 1915, the German Reich, France, the USA, the Austria-Hungarian Empire and other countries had had a gold or bimetal standard, just as England did. The largest and most important countries had even been tied to a gold standard. The Bretton Woods agreement ensured that a certain percentage of the coins and bank notes issued was backed by gold. The price fixing at $35 USD resulted from an American decision taken in 1934 when the U.S. authorities ordained to buy gold at this price. This resulted in a physical gold drain to the USA at the time.

Bretton Woods "35 US-Dollar"
Source: *Christoph Eibl*

The reopening of fixing in 1954 was based on the Pound Sterling, just as before. But since the USA had fixed the gold price at $35 USD, one question would be unavoidable over time: What happens if the dollar is devalued? The question was more than justified, because the wars in Korea (1950-1955) and in Vietnam (1964-1973) did little to fundamentally support the currency.

Due to the rising demand for gold, mostly from investors, a method of easing the buying pressure had to be found. The large national banks of the USA, Great Britain, Belgium, France, Italy, Holland, Switzerland and the Federal Republic of Germany pooled their efforts to keep the gold price at $35 USD by constantly acting as sellers.

As the Vietnam War reached its high point in 1968, the demand for gold outstripped the gold pool of the banks mentioned above to such an extent that the British government, out of the blue, announced a bank holiday for March 15. The gold market remained closed for another two weeks thereafter.

As I mentioned in the beginning, this event changed the gold market radically. The first thing that happened was that the gold market was abruptly freed of all intervention, while the London gold market itself was no longer available to anyone—neither to the central banks nor to the mining industry. The mining industry played a primary role particularly because South Africa usually sold its entire production to London. South Africa was duly irritated and disgruntled and hence very receptive to an offer coming from Switzerland: Switzerland recognized the precariousness of the situation and established the gold pool consisting of the Union Bank of Switzerland (UBS), the Schweizer Bank Verein (Suisse Bank Association - SBV) and the Credit Suisse (CS). These establishments immediately entered into negotiations with the South African mining industry in an endeavor to replace the lapsed London market. The initial consequence for London was the loss of South Africa's mining production. But the loss of confidence London suffered by closing the gold market would linger on for years and took a very long time indeed to restore.

After a two-week break the gold market reopened in a completely new shape. The gold price would from now on be published, traded and fixed in USD. This notation in dollar alone indicates the extent of England's loss of status. But at least the London market had not gone down completely, because the "brand" Loco London was not only used for the fixing, which characterized the physical trade, but also for international standards due to the London Bullion Market Association. The LBMA was established as a trading association by and for the members of the gold market in 1987. It is responsible for coordinating market activities, the formulation of market practices and the definition of a professional etiquette. The LBMA moreover perceives itself as a lobby association intermediating between the markets and regulating authorities.

The LBMA's main objective is to ensure positive guidance and development of the market. To this end, refining standards, the London Good Delivery standard and membership standards have been defined. The LBMA breaks the market participants down into three categories:

1. Market makers (full members) are banks who have entered into the obligation to constantly supply the market with buying and selling prices for certain volumes, types and shapes. They generally enjoy the best reputations, because they are the most experienced market players. Only market makers may usually hold a seat in the fixing.

2. Ordinary members (full members) are banks, trading houses and refineries that actively trade in the gold market and maintain an office in Great Britain. Foreign companies may only become full members if they are equipped with a bank license for England, to quote one condition.

3. "Associate" is the most recently established category and opens the possibility of also accepting companies not resident in the UK into the LBMA circles. These companies are not granted full membership but only given the status of having been screened and accepted into the London market and LBMA circles. International associates are usually smaller banks and gold mines who only trade in the gold market secondarily. The LBMA also pushed the international precious metals trade another step forward by developing official trade statutes in collaboration with the International Swap Dealer Association (ISDA). These statutes are in use around the globe for the spot, forward and options trade.

Logo of the London Bullion Market Association
Source: *London Bullion Market Association*

The LBMA entertains several committees to investigate specific subjects. The physical committee, for example, is responsible for safeguarding and determining the status London Good Delivery. The committee achieves this by licensing foundries and refineries who receive credibility along with their seal of approval. The LBMA has developed an extensive set of regulations for the approval of bars and qualities. London Good Delivery requires a minimum purity of .9950 and a weight of between 350 and 430 ounces. The bars always need to weigh in at a multiple of 0.025 ounces. If they don't, rounding adjustments are made according to the LBMA's physical bullion guidelines. As far as the fineness is concerned, the rounding occurs below the third digit after the decimal. There's a very close interrelationship between fineness and troy ounces, as the following sample calculation shall illustrate:

A bar of 404.075 ounces and a fineness of .9958 has a pure gold weight of 402.378 troy ounces (404.075 x 0.9958 = 402.377885).

It is this high quality that lends the London market its top position internationally. Most international trading houses perform their trades Loco London because they can this way make sure, given the high standards, that their trade's status is relatively safe. No other local market has set itself standards as high as this, which means that Loco London is still a pretty exclusive trademark today.

The London gold fixing is also a survivor. Every attempt to hold fixings in other countries failed or ended in insignificance, respectively. The fixing procedure in its original form has been described above. This original form ended in May 2005, when a new era in fixing dawned.

The banking house NM Rothchild has only very recently announced its withdrawal as a market maker from all gold market activities. It terminated its fixing activity as of May 2004. The announcement came as a shock to many market participants as this very bank had remained loyal to the gold trade for centuries. The remaining members decided that the fixing will from now on be held via telephone, like the other precious metal fixings, and that the chairmanship will rotate annually.

Rothchild is by no means the first bank house to terminate all its gold activities in the London market—Credit Suisse left the market as early as 2002. (Credit Suisse now only operates in the gold market on a reduced scale from Zurich.) But even despite these withdrawals the London gold market remains the center of the entire physical trade, then as now. The significance of the fixing will survive, just as the trade term Loco London will. The long history this market can look back on cannot be simply wished away, and all active market participants around the world are quite aware of this.

UNITED STATES OF AMERICA

The heartbeat of the precious metal trade starts at 8:20 a.m. EST. This is when the Commodity Exchange (COMEX) orders the determination of the first price of the current month of the gold future contract. It is now either 14:20 (Zurich) or 13:20 (London) p.m. in

GOLD

Europe. The morning fixing is over, larger orders from the Middle East have been worked through and the distribution of the physical gold flooding into the Indian subcontinent has been commissioned.

The opening of the New York market starts off an entirely different gold trade that doesn't have that much in common with the European market. This second trading phase is characterized by a lot of noise, a hectic pace, rapid price fluctuations, great speed and large turnovers. And the market participants themselves are just as different as the nature of the market. In London and Zurich, most traders require physical delivery or are at least more closely tied to the physical market, unless they specialize in pure hedging transactions, that is. The American market, on the other hand, is frequently overrun by pure speculators. They find the COMEX more attractive because they can always see the sales volume, which is published via the exchange, and they can more easily and inexpensively extend or shrink their positions due to the COMEX's high and transparent liquidity and the narrow market prices. In contrast to Zurich and London, the USA has a standardized futures market. Although forward contracts (non-standardized futures trades) are largely traded in Europe, the USA has managed, by opening the commodity exchange in 1974, to establish an institution that has become an indispensable part of the gold market.

This chain of events was set in motion in 1974 in the World Trade Center when the commodity futures exchange of the time announced the start of a gold and silver contract. Looking back from today's perspective one can see that there have been several exchanges in North America that also traded gold contracts, but none of them was ever as successful as COMEX.

Canada's Winnipeg Commodity Exchange had been the first to take up institutional exchange trading in 1972. At the time, private gold ownership was still illegal in the USA (as it was from 1934 to 1974 altogether). Although Canada had a healthy mining industry at hand, five American exchanges took up trading in gold contracts two years later. A total of three of these were based in Chicago: the Mid America Commodity Exchange Chicago (MACE), the Chicago Board of Trade (CBOT) and the Chicago Mercantile Exchange (CME/IMM). The other two were in New York: The Commodity Exchange (COMEX) and the New York Mercantile Exchange (NYMEX).

COMEX Floor
Source: COMEX

Over the years, takeovers occurred between individual exchanges or the contracts were terminated as their liquidity and attractiveness was no longer a given. Today, only two exchanges are still active in the gold trade: the NYMEC and the CBOT in New York. The original COMEX merged with NYMEX in 1994 and is now officially called the COMEX Division. But it is still known as the COMEX in market circles and also still called that. The other exchanges fell victim to the consolidation in the gold sector. The CBOT trades a one-kilogram future, the so-called Mini Gold Contract, but this is not very significant in the international market action.

The global gold trade still hinges on the COMEX. There are many reasons for this. The lifting of the gold prohibition in 1974 all at once created a massive demand in a market that had been subjected to restrictions before. As soon as the New York forward exchange had been opened, the mining industry, investors and speculators alike could at least partially economize on overseas transactions and instead gain access to the bullion industry on a local level. It was, due to the active and healthy mining industries in Canada and the United States themselves, highly probable that a futures market would not only attract investors and speculators, but also, and more importantly, large buyers and suppliers. And with this type of market participant a market can generally create a healthy foundation in terms of presence, reputation and liquidity, which is exactly what happened in the USA. Although turnovers grew at a slow pace initially, the market soon managed to attract speculators. In contrast to Europe, American interest in speculation has always been keen. And most interesting of all was the speculation in raw materials, which had always been known to be highly volatile due to various factors like failed harvests or excessively ample extraction volumes. The gold market quickly managed to benefit from these different groups of traders.

But the final breakthrough only came about with the consolidation into "one" exchange. The consolidation had been preceded by a period of searching and development. None of the exchanges knew which contract specifications were most likely to attract the market. The primary question was whether the contracts should be denoted in ounces or kilograms. All the exchanges, except for the Mid-America Commodity Exchange and the Chicago Board of Trade, opted for an ounce quotation ranging from 20 to 200 ounces. Canada wavered for a long time between 100-, 200- and 20-ounce contracts but gave up the gold trade completely in 1988, as the attention and trade volumes increasingly gravitated to the USA.

In the USA today, gold is only traded at CBOT and NYMEX (under the name COMEX Division). The CBOT has kept the trade in one kilogram futures going to this very day. The current disposition of the IMM and the Mid-America Commodity Exchange has either changed completely or they no longer trade in gold contracts.

The finding process took about 15 years. New York, the financial center, finally managed to establish itself internationally with a 100-ounce contract, while the CBOT continues with the one kilogram contract, which is mostly traded by investors and much less in industry.

The COMEX currently trades in 19-month contracts; that is, every two months for the following years with the exception that the front-end months are traded one after the other. The following table shows the current contract terms and prices based on the situation on February 1, 2005. NYMEX even has a separate trading area for the gold futures trade.

Jan 05	426.3 $	Dec 05	435.2 $	Jun 07	463.5 $
Feb 05	424.4 $	Feb 06	440.6 $	Dec 07	472.5 $
Mar 05	426.9 $	Apr 06	443.4 $	Jun 08	481.4 $
Apr 05	426.7 $	Jun 06	446.1 $	Dec 08	490.6 $
Jun 05	429.2 $	Aug 06	448.9 $	Jun 09	499.8 $
Aug 05	431.0 $	Oct 06	451.8 $	Jun 09	509.3 $
Oct 05	435.3 $	Dec 06	454.7 $		

*Current Contracts at COMEX with Reference Price Based on
February 1st, 2005*
Source: *COMEX, own representation*

The area where the actual trading takes place is always called the
"floor." Sometimes the expression "ring" is also used, but nowadays it is only really customary at the London Metals Exchange
(LME), an exchange that specializes in base metals. The desig-
nation is not entirely deliberate as the association with a boxing
ring virtually suggests itself. The gold trade at COMEX is still a
so-called open-outcry trade, a non-electronic type of trading that is
performed verbally, via gesticulation, and bilaterally. The trading
area is separated into several steps that each represent a different
month. A trader standing on the first step, for instance, would trade
the March contract whereas his colleague on the fourth step would
trade the July contract, etc. The front end (the contract closest in
time) is traded in the center of the floor (the whole trading area is
confusingly also called the floor).

The pits are arranged around the ring (or floor). These can be
acquired by individual broking houses. The pits are like their
interfaces to the outside world. This is where the brokers receive
their client orders and verbally communicate them to the traders
standing in the ring. As this procedure can get very noisy indeed
there's always also a sign language facilitating the communication between traders. A stretched-out hand with an outward facing
palm, for instance, means that the person is the seller of a certain
number of contracts over 100 ounces each. Should the trader turn
his hand around he would give the signal for a certain number of
contracts to buy. The illustration on page 76 shows some of these
hand signals.

GOLD

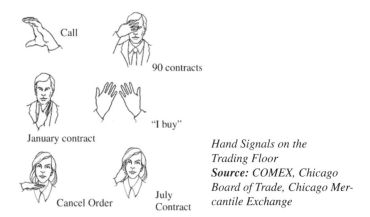

Call

90 contracts

"I buy"

January contract

Cancel Order

July Contract

Hand Signals on the Trading Floor
Source: *COMEX, Chicago Board of Trade, Chicago Mercantile Exchange*

The actual trading goes something like this: Because a whole trading day has happened in Asia and Europe between last night's closing and today's opening of the exchange, there's a high likelihood that the previous closing price and the next day's price in Europe differ, particularly if political or economic events have taken place in the meantime. A new price is therefore determined each morning at 8:30 a.m. EST, which is calculated from the closing price and the price last traded in Europe.

The greatest noise and turnovers are usually achieved right at the start of trading as various stop loss or take profit orders need to be executed. At the start of trading each broker takes stock of his positions—or rather those of his clients—and enters the trading floor to directly execute them. As soon as the opening gong is heard, the traders start shouting their buy and sell orders onto the floor. A deal is closed whenever an opposing party can be found. The whole process is expressed in abbreviations only, unless the traders prefer to exclusively rely on sign language, of course. If a trader says, "five at twenty," for example, he means that he would like to sell five contracts at $452.20 USD (the trader only mentions the digits after the decimal, as everyone is fully aware of those preceding it). The only people to exactly know the market price are therefore the traders, as news services like Reuters or Bloomberg only quote the price last traded. The spread between bid and ask usually ranges from $0.10 to $0.50 USD, but can also be as high as $1 USD. The traders who are standing on the floor usually wait for orders to be transmitted to them by the brokers in the pits.

To a neophyte, the whole trading process often appears to be extremely inefficient and even more prone to mistakes. But interestingly enough, the opposite is actually the case. The error quota is very low, although trading is only conducted verbally or by signals. And the efficiency is also a given, as large positions are actually absorbed by the market.

The floor trading period (4:40) is rather short compared to other markets, but the New York COMEX is further fleshed out by the access system developed by NYEMX. This is the New York Future Exchange's electronic trading platform, which is being used globally from just after closing until shortly before the next morning's opening of the exchange itself and which manages to generate trades and turnovers in the meantime, albeit partly insignificant ones. The online trading tool would hardly suffice as a reference market but it is another liquidity provider rounding off the market in conjunction with the EBS and the trading platforms of banks.

Apart from the classic futures trade, the COMEX also offers option contracts on the futures. This is where fixed term options are traded analogously to the futures market. The important aspect is that the option's underlying is the futures and not the spot market. The options market is smaller than the futures market but still serves the OTC market as a reference. This trade is also conducted in a ring, which, in this case, lies in a separate area.

Futures markets have one peculiarity: Virtually all trades are conducted via cash settlement, meaning that a long position is usually not delivered but balanced via financial payments.

The exchange nonetheless has a sophisticated warehouse system on hand where deliveries resulting from purchases or sales are executed.

Each depot is inspected by the COMEX and, in case of a positive evaluation, added to the list of official warehouses. Similar to the minimum specifications concerning bar size and quality required by the LBMA, the COMEX also has certain conditions: Deliveries in gold need to have a minimum purity of .9950. Bars need to have a minimum weight of 100 ounces and their total weight needs to be a multiple of this. As regards authorized refineries or bar producers, respectively, the COMEX quite closely copies the LBMA.

The various depots operate in the classical sense with warehouse receipts, i.e., if a long futures position changes hands, the buyer is provided with a warehouse receipt granting him a claim to the gold's delivery.

As we can see, both COMEX and LBMA have set themselves very high standards concerning the handling of physical gold. But an actual physical execution is rarely required because only about one percent of all sales is either notified for delivery during the notice period or expires without having been settled. This fact graphically illustrates the primary interest of market traders who frequently only wish to participate in price fluctuations. There is often no connection to the physical trade at all, and this is also one of the New York market's main characteristics. The COMEX is a meeting point for speculators, large funds and CTAs. And in a period where hedge funds and alternative investments have become a daily occurrence, the importance of the New York market place is all the more clearly highlighted.

BRAZIL

Most gold traders would predominantly associate Brazil with gold mining and to a much lesser extent with a regulated spot and futures exchange. But although it is true that Brazil's mines are more important than the country's international trade, the Bolsa de Mercadorias e Futuros (BM&F) was established in Sao Paulo as early as 1985 and held its first gold trading session in 1986.

Bolsa Mercardorias e Futuros
Source: *Gold Refiners & Bars Worldwide*

During the 20th century, Brazil constantly experienced inflationary problems. Gold almost became a stock-in-trade of a population keen on hedging, leading to a continuous demand from that side. But the mines producing gold locally either did not have access to a functioning OTC market or were unable to rely on a respected or creditable OTC market. Along with the efforts of the BM&F, a gold association was established in the same year, the Associacao National do Ouro (ANORO) whose members include financial institutions, mines and consignment companies. The ANORO was intended to establish a standardized Brazilian market along the lines of Hong Kong or London. It was decreed that all companies wishing to trade at the BM&F first had to be approved by ANORO. In addition to this, a partnership was entered into between ANORO, the Central Bank of Brazil and the BM&F, opening up the possibility of providing the institutionalized gold trade with a truly stable foundation. The trade currently includes spot contracts, gold futures and gold options. The gold spot contract refers to a 250-gram gold bar with a minimum purity of .9990. The price is expressed in Brazilian Rias. The bars furthermore need to come from one of the refineries accredited by the BM&F. The futures contract includes a limit up/down, just like the TOCOM, whose fluctuations are not permitted to exceed a price range of 5 percent of the previous day's closing price in either direction. The term "limit up/down" refers to the fluctuations limit that may be traded maximally in a day and is prescribed by the exchange. An exchange with this type of contract will therefore never suffer from drastic price drops or strong price increases. The contracts are similar to those traded at COMEX.

Internationally, the BM&F has little importance in the gold trade. But it is interesting to see how exchanges can sometimes spring up in some far-off places and still function, predominantly based on local businesses. It is possible, under certain circumstances, for COMEX and BM&F traders to transact a classic Loco arbitrage, as each markets exposed to different influences: The COMEX comes under the sway of speculative buyers and sellers, whereas the BM&F is more strongly influenced by physical supplies from the mines.

TURKEY

Turkey is often regarded as the door to the East. That sounds more miraculous and mysterious than it actually is. Since the beginning of the 20th century and the government of Kemal Attatürk, Turkey has actively been engaged in Westernization. But even despite this early liberal movement, the gold market was only reformed in the 1980s. The first gold transactions were based on a gold index applying to Turkish Lira.

But by the end of the 80s, gold was also traded in foreign denominations. The Istanbul Gold Exchange (IGE) in its current form, however, was only established in July 1995, and option trading started the year after that. Trading is divided into two sessions, as is the case with TOCOM in Japan. The early trading session is from 11:00 a.m. to 1:00 p.m. and the afternoon trade takes place between 3:00 p.m. and 4:00 p.m. The trading times were chosen in order to have a reference market in the London market, which runs parallel.

The IGE has a cash market as well as a futures market coupled to option contracts. But the Turkish spot market functions differently to the Loco London market. In the regulated gold market, the standard transactions are traded as gold finenesses of .9950 to .9999. Any gold bar can be delivered, ranging from just one to 1000 grams, as well as the 350- to 430-ounce bars representing the LBMA's standard bar. Transactions can be denominated in Turkish Lira, USD or Euro. Depending on the currency, different payment terms are binding. Turkish Lira and USD payments may be delayed up to (t+5), or five days, whereas a Euro payment only permits (t+1), or one day. Compensation payments for differences between the traded exchange rate and the one on the valuation date are collected in the shape of a clearing fee. The IGE generally has few requirements regarding minimum amounts. This fact highlights the exchange's primary client potential and shows that the existing exchange structure does not address the global inter-bank market.

Apart from the standard transactions, there are also non-standard transactions permitting gold of lesser quality to be traded. The spot trade is supplemented by future and option contracts. There are two contracts here, as well: one on 100 ounces and the other on one kilogram of gold with a minimum purity of .9950. The price is

quoted in USD per ounce and Turkish Lira per gram.
As is customary in all futures exchanges, trade in Turkey functions
via margin payments.

Trading, or the execution of orders, can happen in two ways. There
is the floor trade and then there is the possibility to trade electronically. The latter functions via an electronic matchbook connecting
prices and incoming orders logarithmically. There is insofar little
difference to other futures exchanges. Trading in futures and options contracts is possible between 11:00 a.m. and 4:00 p.m.

But that's not all: The IGE also has an official lending platform.
This product is in this shape unique in the world and therefore
serves to distinguish this exchange from all other large exchanges.
The platform's purposes are the evaluation of gold lending demand
and to manage everything as efficiently as possible via one center.
The primary intention is certainly to provide a platform for smaller
jewelry producers and wholesalers.

Turkey is more of a classic, physically oriented, market than a pure
gold investment market or an inter-bank market. This is why there
is also, in addition to the trade in bars, a lively scrap trade, meaning
the market for so-called old gold or even recycled gold. Another
important aspect is that the Turkish population is brimming over
with avid gold consumers, which ensures a flourishing jewelry
industry. But the buyers for this industry predominantly demand
14- to 18-carat gold, as opposed to the higher purity 22-carat gold,
as jewelry production is a lot more profitable this way.

Turkey plays a dual role. The country has on the one hand ap-
plied for membership in the European Union, but on the other is
strongly characterized by Arabic influences in its eastern regions.
This makes Turkey an important hub for the gold trade between
Europe and the Middle East, at least in terms of the physical trade.
Physical gold deliveries are a daily occurrence throughout Turkey
because there is lively consignment traffic from the gold trading
center Istanbul to the regions bordering the Middle East or the
Caucasian republics.

MIDDLE EAST

The Middle East is comprised of numerous countries, many of them quite small. There are several cities and states in the region that are involved in the gold trade, but Dubai, Beirut and Kuwait City are of the greatest importance.

Dubai, located in the United Arab Emirates, has a relatively short history, as the local gold trade only gained any significant importance in the middle of the last century. A wag would characterize the gold trade in Dubai with one word: India.

India is crucial for Dubai and its towns in the gulf region. And Dubai offers one great advantage compared to many Indian ports (and other ports in the region) by being a free port and therefore not charging levies like gold duties or taxes. In addition, the general prosperity of the region's oil producing countries provides the large amount of capital required to pre-finance gold that is to be sold on to India, as was the case during the end of the 1960s and beginning of the 1970s, when traders in Dubai bought up virtually all the demand there was to come from India on the English and Swiss markets. Because Indian demand primarily consists of small bars (T.T. bars, kilogram bars and gram bars), suitable refineries had to be found who commanded the appropriate standing and experience concerning these products. Initially, British refiners Johnson Matthey provided most of the bars, but later the Swiss refineries greatly increased their share. Today, the Swiss PAMP SA is a very important refinery and simultaneously the largest provider of small bars to the Indian subcontinent.

The trade in Dubai is a classic consignment stock trade. That means local banks such as the Standard Chartered Bank Dubai entertain local custodian warehouses, which are financed by British and Swiss banks. Custodian warehouses are safekeeping contracts between the owner and the custodian. The USB Zurich can hence deliver, on its own account, gold to Dubai into the custodian warehouse of the Standard Chartered Bank Dubai's custodian. The Dubai bank now has custodianship of the gold, but doesn't own it outright. The legal separation of asset ownership is also clearly demonstrated in a physical separation. The Standard Chartered Bank Dubai can always, should it require the gold, request a quote from UBS in Zurich and, if it decides to buy the gold from the

UBS, have it simply delivered from the custodian warehouse's vaults. In this manner, the Standard Chartered Bank Dubai procures financing, as it is not required to hold the precious metal, but can always fall back upon it at short notice. The local banks in Dubai, Abu Dhabi, Bahrain or Kuwait City deliver this gold directly to India. (The bank names used here only serve to illustrate the example—the actual number of active bank houses is considerably greater.)

Dubai also serves as a transfer point for directly extracted South African gold. In geographical terms, Dubai is a lot closer to the Indian ports and airports than either London or Zurich. The gold, which has already been processed into marketable bars in South Africa, can in this manner be directly sold to India via Dubai. But South Africa has just two established refineries that can only supply a small fraction compared to Switzerland, Germany and Great Britain. One is the world-famous Rand Refinery, which achieved its iconic status through the long-term production of the bullion coin Krugerrand, and the other is the refinery Musuku of the gold mining company Harmony.

Kuwait City, Bahrain and Abu Dhabi are also tied to the Indian trade in a similar manner to Dubai, but they are less important. Local traders nonetheless distribute a relatively large amount of gold to the jewelry industry, which has been growing disproportionately due to the ever-rising oil proceeds.

Up to the beginning of the civil war in the 1970s, Beirut in Lebanon was one of the main trading places for gold that was to be distributed into Middle Eastern channels. Because the changed conditions did not allow for a local gold trade, most traders left the country, and the Lebanese gold trade lost its significance entirely during the 70s.

The region itself is regarded as very significant, especially now because the government of the United Arab Emirates has been working toward the establishment of a commodity exchange for a number of years. These plans were incrementally realized in recent years and the Dubai Metals Commodity Center (DMCC) will open in Dubai this year (the year of publication).

The greatest importance is attached to energy and gold here—the first because it is the main source of income for the countries mentioned and the latter because it serves as an alternative to oil sales with the USD proceeds. To this end, the DMCC has summoned international heavyweights to Dubai in order to develop and extend the trade. It remains to be seen whether Dubai will be able to strengthen its position even further.

SWITZERLAND

The Alpine republic has two cities that are equally well-known but actually radically different: Geneva and Zurich.

Let me start with a little historical background. Switzerland has always been a neutral country, a nation with a very strict banking secrecy that attracts capital from around the globe. No great importance is attached to the question of where the money comes from and never has been. This has often resulted in Switzerland becoming embroiled in various controversial discussions. During the Second World War, for instance, a large amount of German gold was sold to Switzerland. This at first gave little cause for ethical concern, until the sales in 1943 significantly exceeded the volume of gold stocks extant in the Third Reich before the war. It became plain as daylight that the Reichsbank was selling gold it hadn't originally owned but had instead looted from other central banks (for instance in Belgium and Holland). This and more importantly the fact of Switzerland's complicity resulted in international discussions that damaged the country's reputation.

But Switzerland had been involved in the gold trade much earlier than that, as the country's relevant history can be traced back well over 100 years, the main reason being that the nation was one of the last to abolish the gold standard in the year 1936.

The affinity to gold had hence always been there and the Swiss National Bank therefore always had some business to conduct with local banks. Swiss banks were the first buyers to be active in the London gold fixing in 1919, for instance.

Switzerland had a general demand for physical gold because the country harbors three internationally renowned refineries (a fourth was added later), but no gold production of its own. The main problem with this was that, from the 1950s to the 1970s, approximately three-quarters of global annual gold production originated from South Africa, but that the South Africans had direct contracts with the London banks. This meant that the lion's share of the production went to England.

An additional factor was the engagement of the Soviet Union who primarily sold its gold stocks via London in the beginning of the 1960s. As a result, the Swiss were the biggest buyers at the London fixing between 1954 (the year it reopened) and 1968. On the one hand, Swiss private banks had great demand for physical bars (predominantly for clients who had their assets managed in Switzerland), and on the other, the Swiss refineries needed gold to maintain their own productivity. The infrastructure was there; gold ownership was tax exempt in Switzerland and the refineries distributed their bars around the world, thereby gaining a very good reputation. The Swiss banking network, now operating worldwide, functioned perfectly.

But only after the gold pool collapsed in 1968 was Switzerland able to reroute the South Africa-London distribution channel directly to Zurich. South African irritation was quickly mollified by the Swiss banks, which had started holding preliminary talks with the South African authorities long before this step was taken.

Occasioned by the 1968 gold crisis, the three largest banking houses—the Union Bank of Switzerland (UBS), the Schweizer Bank Verein (SBV) and Credit Suisse (CS)—the decided to coordinate their activities to a certain extent. The result came to be known as the Zurich gold pool. Each of the three banks retained their independence and acted as independent market makers. And because the three banks were closely associated to refineries, the collaboration was extended into this area, as well. In 1982, these activities finally resulted in a common broker, the Premax. Partnership interest was evenly distributed among the three banks. The broker, who currently still enjoys a good reputation for gold forward brokerage, took over quotation and therewith coordinated the trade.

Even in the 70s, much more than 80 percent of South Africa's production was traded via Zurich's Bahnhofstrasse where the most important Swiss banks are located. But Zurich also had another windfall besides South African gold.

GOLD

In 1972, the Soviet central bank decided to execute major gold sales via Switzerland. Statistics show that the former Soviet Union sold over 2,000 tons of gold via the Swiss market in the years following 1972.

The trade naturally involved the larger as well as smaller banks— banks like Julius Bär or the Bank Leu, which still shines today by virtue of its Münzkabinett (coin cabinet). On the industrial side, the refineries Metalor SA in Neuchâtel, as well as Valcambie SA and Argor SA, both in Chiasso, produced bars that were in demand globally. The Metalor SA is primarily known for its close affiliation to the clock market and its products for the dental industry, whereas Argor SA's main focus was bar production, and still is today. The reasons for this were simple know-how and the proximity to the Italian jewelry market. In 1986, Argor SA entered into a strategic partnership with the German company W.C. Heraeus in Hanau and is now known as Argor-Heraeus SA Valcambie, part of Credit Suisse.

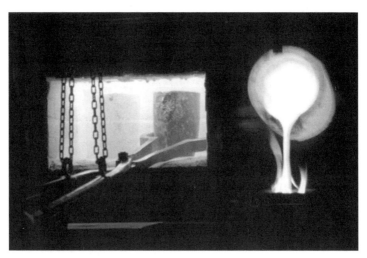

Interior of a Refinery
Source: *Gerald Hoberman*

After Credit Suisse retired from market making and withdrew as an important gold trader from London, this company took over second place.

In the mid-70s the long-standing local refineries came up against some serious competition in the shape of PAMP SA, or Produits Artistique de Metaux Precieux. The company had been established by the Lebanese gold trading family Shakarchi who had moved to Switzerland because of the civil war in their home country. The PAMP SA is part of MKS SA, which is probably the global physical gold trade's largest private and bank-independent company. In the market, MKS SA is famed for its global distribution channels which make the company an excellent trading partner for location swaps. And the 100 percent subsidiary PAMP SA opens up the possibility of offering quality and size swaps. MKS SA satisfies a large part of Far Eastern gold demand and is an important market player concerning India's demand for physical gold.

We can already see, without once mentioning Geneva, that Switzerland is of central importance in global gold trading. As far as Zurich is concerned, it is important to mention that Zurich has a Loco trade. This means that gold trading is also possible globally with delivery in Zurich, parallel to the Loco London trade. The local clearing is mainly done by the UBS who became the most active Swiss bank by far after Credit Suisse withdrew in 2002, and who is of primary importance today. Although Credit Suisse never withdrew entirely from the gold market its significance is much diminished.

The Geneva story is different and much shorter. Geneva doesn't and never did have any significant industrial clientele, which was mainly served by Zurich instead. The classic Genevan client is a rich private businessman, an asset management company or a wealthy family with an international background who has its assets managed by a Swiss private bank. In the Geneva market, this results in a large demand for spot trades that are more often than not also delivered physically. As far as gold account trades are concerned the bars usually remain at the Zurich market place and are only posted electronically.

Geneva is where the "golden" high-finance resides. And gold is as much a part of this city as the lake and the Alps. In the gold market, Geneva stands for an innovation that was first developed by Valeurs White Weld in 1976: OTC gold options. Their development and market introduction was clearly initiated by Geneva and relatively quickly adopted by other gold trading locations. Later the trade in gold-option contracts also became available in New York at COMEX and throughout the entire inter-bank trade.

Not that much has changed in Switzerland in 2005. The USB and its team still internationally trade from Zurich and remain an active market maker in London. And Geneva is still the home turf of large asset managers who purchase gold in various shapes and via various methods on account of their clients.

LUXEMBOURG

As a gold trading location, Luxembourg has a very special role. The country was turned into the center of the continental European gold trade in the 1980s by various banks and trading houses. The reason was the decision taken by the Swiss and German governments to levy VAT on gold acquisitions. And as if this weren't enough, the French government stopped the anonymous purchase of gold. A new refuge was found in Luxembourg and the capital flow in shape of bars, Krugerrand and gold certificates immediately started heading for neighboring Germany.

Compared to that period, one now has to say that Luxembourg has almost waned into insignificance as a trading place. It certainly still benefits from its offshore status but as far as gold is concerned, there is only one local bank with a significant trading volume left in the Luxembourg market: The Commerzbank Luxembourg SA.

The Commerzbank AG transferred its entire gold trading team to Luxembourg fairly early on, primarily in order to benefit from the tax differences in Europe's center.

Luxembourg's own gold fixing was established on March 17, 1981 (VAT on gold had been abolished in 1979) and always published the first official morning price in advance of the London gold fixing. The development had been preceded by a budget law which lifted VAT from all gold transactions not serving industrial demand as of January 1, 1978. Although Commerzbank Luxembourg SA currently occupies pride of place, the bank had not been the initiator of the first fixing, which was started by local banks like the Kreditbank SA Luxembourg SA instead.

The fixing was based on a one-kilogram bar. This fact by itself shows clearly who participated in the country's market: the kilogram as the unit of measure of continental Europeans and the one-kilogram bar as a classic retail investment product. The market was primarily made up of private investors and asset managers.

Fixing takes place at 10:30 a.m. and concerns one kilogram of gold with a fineness of .9999, to be delivered in standard bars. In the past, the price was fixed for Loco Luxembourg. The fixing of a 400-ounce bar was sufficient to compliantly fix the price.

In the late 1980s, when the market had been firmly established, also for local banks, the classic bullion trade followed (i.e. trading and hedging involving the mining industry and central banks). The Commerzbank Luxembourg SA, the "yellow company," was able to increase its local dominance by entering into a partnership with the refinery Argor, located in Switzerland.

Today, the fixing has all but lost its significance for international trading. Looking back, one must concede that it could never measure up in terms of border-crossing success stories—except for local traders. But the Luxembourg market has, ever since its inception, concentrated on building structures, and, to this end, established a local association called Golddealers Luxembourg which was able to attract approximately 100 members in a very short period of time. This association existed for approximately 10 years.

From today's perspective, the main reason for the market's establishment was simply the fact that it offered the advantage of VAT-exempt trading. As soon as VAT-on-gold acquisitions was abolished again in Germany and Switzerland in the 1990s, the capital inflow from these countries abruptly came to a halt. Most important was the abolishment of VAT on the so-called investment bars (kilogram bars and 100-gram bars) and bullion coins (the most famous bullion coin being the South African Krugerrand). Examining Luxembourg today, one soon discovers that the Commerzbank Luxembourg SA still keeps the flags flying, but also that it alone represents just about the whole market. Gold investments are certainly still being placed with the remaining asset managers here and there, but the market nowhere near benefits from fresh capital inflows on the same scale as before. The only remaining advantage Luxembourg has to offer now is principally its offshore status.

FRANCE

France plays a leading role in the gold market. The European republic has the third largest gold reserves worldwide: over 3,000 tons. Neither is this situation exactly novel; in the 1960s, France even held the top position globally.

The direct impact on the gold market and trading activities is primarily that the large gold holdings needed to be managed somehow. A job that was mainly undertaken by local Paris banks, who have retained their structure to this day like the Crédit Lyonais or the Société Générale.

In addition to their own client trade, these banks also traditionally engaged in executing the central bank's market activities. In recent decades the central bank variously acted as a buyer or seller—and that's not counting the large volume of lending and forward business. The result: The local banks managed to gather the necessary knowledge and created their own structures in the Paris banking trade. The whole thing was a success because Paris had had its own fixing in recent decades.

Until very recently, the Paris gold fixing was held daily at 12:30 p.m. Only in 2004 was it finally abolished due to low turnovers and general insignificance.

As is the case with the changes affecting London gold fixing, the international gold market lost a further part of its tradition with the abandonment of the Paris fixing in 2004, even though this fixing had only been of secondary importance. The end of the fixing also effectively ended the small inter-bank market itself, as there now is—besides the large Société Générale, which is active as a market maker in London—only the CPR Or Trading left, which is mainly responsible for conducting the local gold trade. But talks about the renewed introduction of a gold market place in Paris are apparently underway with Euronext, the Franco-European exchange. But because London is Europe's leading gold market and won't easily put up with competitors, the demand for a Paris gold market is limited among both buyers and sellers.

The French gold market has been subjected to more restrictions and monitoring regulations than any other European market. In

1968, it was still free of any tax and monitoring mechanisms. But that rapidly changed and the subsequent period until 1987 was characterized by import and export limitations. Gold sales were also made taxable. The 7-percent tax might have been a bargain compared to the German taxes of the time at 14 percent, but in comparison with the entirely tax-exempt Luxembourg market, Paris was nonetheless not very attractive for gold investors. For this reason a veritable capital drain to the East was underway, running parallel to the smuggle between France and Luxembourg.

As if to further impede gold purchasing in France, the newly elected government in the 1980s demanded the abolishment of anonymity for gold buyers. This was, in addition to the export and import restrictions and tax increases, the last inducement needed to drive the gold market underground or across the borders to foreign shores.

The gold fixing managed to survive these difficult times. Compared to the other fixings, the French variant had always been marked by a highly individual style. Apart from the large fixing of the Good Delivery 12.5-kilogram bar or 400-ounce bar, respectively, there also was the one-kilogram fixing and the fixing of France's most popular bullion coin, the Mexican 50 Peso Centenario. During the 1990s, the fixing included nine members (all of them French banks), whose representatives held a daily meeting in the Paris stock exchange building to fix the three media. The procedure was similar to the London fixing. All the members had small, individual cabins equipped with telephones, which were used to communicate with clients or the member's own trading center.

The fixing is no longer extant. French banks are hardly active from Paris nowadays. From a global point of view, only the Société Générale still has any importance due to being a member of the fixing at London's gold market. The bank even moved its gold trade to London as of New Year 2004/2005. The abolishment of the Paris fixing hence represents more of a loss of tradition than of market shares. Today, Paris still has a relatively active trade in gold coins and investment bars, but in global terms, it is insignificant.

GERMANY

To cut to the chase: Germany might no longer play a significant part as a gold trading location internationally, but its market still has a lot to offer. Germany, for one thing, is the home of several important banks and refineries that have managed to build an international reputation over the years.

Analogous to the Swiss and English foundries, Germany features two of the world's best known refineries: the Degussa (now Unicore N.V.) and WC Heraeus, both located in Hanau near Frankfurt/Main. Both Heraeus and Degussa, which has meanwhile become part of the Belgian refinery conglomerate Unicore N.V., have been active in the refining business ever since the middle of the 19th century. Both companies' bars have international status (as both refineries are honored by the LBMA) and both enjoy good standing among the market's industrialists and investors. Both gold bars are undoubtedly available globally, even in far-flung locations. In Germany, their reputations are so unshakeable that bars from anywhere else frequently have a hard time finding a market (and are sometimes only bought up at a discount). Heraeus expanded in the 1990s and holds shares in the Swiss refinery Argor, which is now called Argor-Heraeus and operates from Mendrisio in Switzerland.

It is no longer evident that the refineries in Germany were also primarily required for local commerce because the capital of German jewelry production, Pforzheim, has meanwhile lost significant market shares to countries with lower overheads, for instance in the Far East. The historic situation couldn't have been more different. Pforzheim was the main transfer point for German industrial gold and the town used to produce a significantly larger amount of gold jewelry than today. There are still several medium-sized refineries to be found locally, as well as a number of small-scale operations, with the Allgemeine Gold- und Silberscheideanstalt being somewhat better known due to now being integrated in the Umicore N.V. combine.

What else is there to say about gold in Germany? Quite a lot, actually, as the German Bundesbank holds the world's second largest reserves with over 3,400 tons of gold. Although the Bundesbank, as far as gold sales are concerned, has in recent years acted nowhere near as aggressively as, for instance, the Bank of England or the Schweizerische National Bank, the German central bank still now and again becomes active in the market by engaging in lending or forward transactions. The Bundesbank has also been firmly earmarked as a disposer of gold reserves, and it is to be expected that the institution will sell several hundred tons pursuant to the II. Washington Agreement on Gold.

As a trading location, Germany itself is of little international importance, although there used to be gold fixing until 2001. The fixing used to be officially integrated in the workflow of the Deutsche Börse, i.e., the German exchange. The official exchange broker would contact Frankfurt's largest banks via telephone in order to determine prices for a 12.5-kilogram bar and a one-kilogram bar by around 12 noon.

The two largest German banks, the Deutsche Bank AG and the Dresdner Bank AG, are nonetheless of great importance internationally.

Both banking houses are active in the market as market makers, as is the Deutsche Bank AG in terms of the LBMA. The strength of both banks is based on many years of intensive contact to mines as well as the central banks.

The Dresdner Bank AG, for instance, was commissioned to dispose of the former GDR's gold reserves in the market in the wake of the country's reunification and was the largest buyer at IMF gold auctions in the 1970s. In those days, the Dresdner Bank AG would sometimes absorb over 90 percent of the gold being auctioned. This reputation has stayed with the bank to this very day.

A comparison of German banks with their Swiss colleagues shows that the Swiss usually needed to buy from refineries or enter into strategic shareholdings, respectively, in order to increase the efficiency of their gold trade. The German banks obviously had it much easier in this respect, as the collaboration between the major banks and the local gold industry, as represented by the Degussa or, respectively, Heraeus, which apparently functioned very well, even without cross-shareholdings. In Germany, the distribution of bars was hence mainly limited to the two German brands, a fact that naturally still makes it extremely difficult for foreign bar producers to gain a foothold in Germany.

But similar to France, the German gold market was weighed down heavily by VAT. The acquisition of gold bars had been taxable for a long time. In the 1980s, even the bullion coins became subject to tax levies. The VAT on all gold bars and coins was finally lifted in the year 1993. But by then, a lot of capital had been deposited beyond the reach of German law, to a large part in Luxembourg.

One of the reasons given for the creation of the German fixing had been that VAT could be avoided via the inter-bank trade. So when the VAT was abolished in 1993, the fixing lost this advantage. It nonetheless carried on until 2001, but there were few transactions and even if there were, they were mostly low volume.

Germany still mainly feeds on its reputation in the gold trade and on the achievements of previous decades. By now, both banking houses—the Deutsche as well as the Dresdner Bank—had transferred their trading units to London, New York and Singapore. But the names Degussa, Dresdner, Deutsche and Heraeus still keep the flag of Germany's reputation flying internationally, as the market in general would be a lot poorer in class, tradition and quality without these four names.

Degussa Gold Bars
Source: Degussa

INDIA

The Indian subcontinent is the largest buyer of physical gold worldwide. The demand primarily comes from the jewelry industry. With its 600 tons of annual demand for gold, India is also the most important supporter of the mining industry.

These facts may surprise some readers as India, at least the way it is usually represented in the media, is one of the world's poorer countries. But the large demand for gold is rooted in the nation's history and traditions. Gold is principally regarded as a symbol of wealth, just as it is in many other countries. But in India, gold and gold jewelry are much more frequently given away as gifts or strategically hoarded than in Europe or the United States. It is a local tradition, for instance, to give gold presents at weddings. As soon as a daughter is born, Indian family fathers start hoarding gold bought at regular intervals in preparation of the gold dowry to be paid later.

The wedding itself, an event of great importance in Indian culture, is traditionally held shortly after the Diwali Festival of Lights. The Diwali season starts in October and continues into November, while the subsequent wedding season runs well into February. The exact dates vary from year to year because they are determined according to the Hindu calendar and not the calendar system as we know it. The standard gold purity is 22- to 24-carat and therefore higher than in other countries. Almost half of all gold bought in India is turned into wedding presents, and there are approximately eight million weddings a year.

The demand is hence very seasonal and repeats the same cycle annually. One should also mention, though, that the extent of the demand depends on the harvest or the monsoon season. If the monsoon brought floods and the concomitant damage, the demand for gold is traditionally lower. These happpenstances are therefore also reflected, quite logically, in demand and supply volumes, respectively.

But let us now examine the trading and the character of the market. Indians are basically very price sensitive, meaning that they usually buy more in weak markets and tend to withdraw their buy orders in strongly rising markets. The historic unit of measure in trading is the traditional Tola. The main gold trading sizes used to be the traditional five Tola and ten Tola bars. The latter bars are also sometimes called Bisquits.

GOLD

The situation only changed in 2003 when the Indian government lowered the tax rates applicable to bars denominated in grams.

Due to the aforementioned price sensitivity, the kilogram bars therefore only managed to become customary in recent years. Today's standard is the one-kilogram bar with a purity of .9990. Another Indian standard seems to be constant changes in the law and tax regulations. Past and present examples include fluctuations in gold taxes and levies from district to district. Initially, during the 1980s and 1990s, gold could only be imported via authorized state channels. These regulations were amended regularly, for instance with the Non-Resident Indian law (NRI), which came into force in 1992. As the name says, this law permits individuals who are not resident in India to import gold to the country—only in small volumes, of course (currently up to 10 kilograms). This method is nonetheless being employed by some market participants, sometimes in a commercially organized manner, to satisfy demand in times of high buying interest.

Since the end of British Colonial rule in 1947, the Indian government has kept control of the gold market by regulating not only gold imports and exports, its distribution, production and sale, but also the ownership of the precious metal. From 1963 until 1990, the Gold Act limited gold ownership to jewelry. To own bars of any shape or size was plainly illegal. Institutions and individuals dealing with gold professionally required an authorization. This quite officially served to create a standstill in the gold trade. The law was finally amended in 1990. The government became aware of the precarious situation it had created by the regulation and tried to strengthen its local jewelry industry by subsidizing exports. And the banks were simultaneously encouraged to engage in the gold trade more actively.

But the restrictions of the past had created a very intensive black market and triggered massive smuggling activities, as the local gold price charged to Indian end buyers is usually higher than the international price. This price spread results from transport costs, import duties, trade margins and sales levies. The publicly known wholesale transfer points for the Indian gold market include Dubai, Singapore and Hong Kong. But quite apart from these centers there is also some gold smuggling activity from Afghanistan to India.

The major Indian gold traders comprise about 20 banks authorized to import the metal by the Reserve Bank of India, as well as a number of large trading establishments. These are all strictly bound by legal regulations. This means, for instance, that local bank clients were, until quite recently, not permitted to execute OTC forward contracts with banks. This situation only changed some months ago when the largest Indian private bank, the ICICI Bank, traded the very first OTC forward contract. The efforts to create an Indian market along international standards are currently quite noticeable. India features a total of 19 commodity exchanges, three of which presently trade in gold: the Multi-Commodity Exchange (MCX), the National Multi-Commodity Exchange of India Ltd. (MSEIL) and the National Commodity & Derivatives Exchange Ltd. (NC-DEX). These exchanges are still quite young and it is hard to tell at the moment whether they will be successful or not. Up to now, cash products and futures are traded there. The exchange clientele largely consists of some of the local banks and trading houses. But the remaining exchanges are currently also thinking about offering gold products, as there is a strong indication that the state would like to lift all restrictions one by one and open up to the international market.

There are still some fundamental differences to consider in gold refining and processing. In contrast to Europe, which has four or five major refineries to process virtually the entire prill capacity, India has over 10,000 smaller ones, usually two-man operations. They buy old jewelry, melt it down and then turn it into new jewelry—like a type of backyard-refinery. But they do distribute the bulk of the gold to end consumers. To stabilize the gold market, the Bombay Bullion Association, which is intended to represent the market interests of all participants, had already been established in the Indian republic's early years. But it is only of local significance and does not set international standards, as for instance the LBMA does.

In conclusion, a brief summary: The gold market in its current form would not exist without India. Indian annual demand for gold is approximately 600 tons, and the annual global mine production is roughly 2,500 tons. All internationally active banks and most of the refineries are directly involved in the Indian gold trade, and their earnings to a large part depend—directly and indirectly—on this trade. If deregulation of the local market continues and the

Indian banks start increasing their influence on the gold trade, the European bullion market will have to concede some of its potential sooner or later. But it is hard to predict how large a portion of the market will transfer to India, as the European market can offer two distinct advantages: structures that have developed over time and many years of global experience.

Backyard Refinery in India
Source: *An Introduction to the Indian Gold Market*

SOUTH AFRICA

Any detailed examination of the gold market inevitably needs to mention South Africa. South Africa and gold harmonize like the USA and Coca-Cola. The country's populace is quite dependent on the mining industry, which has been the largest employer for decades.

The whole development started in 1886 in the north-eastern region of the Southern Cape. The first gold discovery was made in the spot where Johannesburg stands today, which makes Johannesburg a classic gold-digging town, although it does not much look like one today. Instead, the gold mining industry is rather unobtrusive. Because whereas open pit mines remodel entire landscapes, mine shafts are much less conspicuous. The mining activity is mainly located underground. South Africa has a uniquely complex and multi-faceted geology. Millions of years ago, magma erosions pushed metalliferous layers up from deep within the earth's interior. These gold deposits nonetheless still lie at a depth of 1,000 to 2,000 meters. This means that some of the mineshafts reach a depth of up to 3,400 meters (for instance in the mining region Western Deeps). Deep drilling (deep mining), has therefore been practiced in South Africa for over 100 years. This is probably the most difficult and dangerous of mining activities and requires highly sophisticated technical equipment.

The gold produced in South Africa has always been the subject of tight restrictions. Even in times of Apartheid the gold mines were not permitted to freely put their gold on the international market. Until recent times, this privilege was reserved solely for the South African Reserve Bank, i.e., South Africa's central bank. The Doré bars produced by the mines had to be directly delivered to the Rand Refinery, which then proceeded to turn them into London Good Delivery bars. And the Rand Refineries were, in turn, obliged to deliver the metal to the South African Reserve Bank which then either put it into reserve or sold it. The gold mining companies were tied to Rand, as it used to be South Africa's only refinery for many years. The authorization has been extended to others in recent times. The gold mining establishment Harmony Gold, for instance, has built its own refinery called Musuku, but it does not fulfill the LBMA's standards and requirements and therefore does not qualify for London Good Delivery status.

Apart from gold bars in all shapes and sizes, South Africa is predominantly known as the producer of the globally most significant bullion coin called Krugerrand. The coin, whose name derives from the Boer King Ohm Krüger, was issued under license to South Africa's state mint, the Pretoria Mint.

Underground Mining in South Africa
Source: *The Hong Kong Gold Market*

The South African trade is closely linked to the trading centers London, Zurich, Dubai and Singapore. Trading houses frequently buy unrefined Doré bars directly from the mines and then fly them to England or Switzerland where they are purified and turned into small bars or London Good Delivery bars. Most of these are subsequently sold to the Indian market. The London and Zurich banks partly buy up the production and have it flown to Dubai or Singapore in order to have it turned into standard bars by the local refineries there. The South African mines hence sell their gold at a discount, which usually ranges between $0.20 and $0.50 USD per ounce. The gold is then transferred to the new buyer who is liable for transport costs and risks—a fact that justifies the discount.

Doré bars
Source: *Rand Refinery*

Fundamentally, the South African banks and refineries compete with the London market. The Standard Bank of South Africa or the Rand Merchant Bank, for example, act as buyers to have this production then refined by refineries such as the Rand Refinery. Whereas the trading establishments in London or Zurich are naturally keen on having the gold purified by one of their affiliated refineries and are hence prepared to aggressively bid for it, so the refining procedure may be continued by local refineries uninterruptedly.

South Africa itself is not very interesting from an investor's perspective, as gold bar ownership is illegal there. Private individuals are only permitted to hold gold coins like the Krugerrand or historical gold coinage. This virtually forces the South African gold market to focus on industry, with the result that there is less of a local inter-bank market. Despite massive difficulties due to the current strength of the Rand in comparison with the USD, high unemployment rates and an immense HIV and Aids problem, South Africa continues to play an important role in the international physical bullion trade. The Krugerrand, then as now, is the most important bullion coin internationally, and thus virtually guarantees that South Africa, the Rand Refinery and its coin production will forever be part of the international gold community.

ITALY

Italy is a typical example of a jewelry producing nation. The country imports gold in the form of grains or bars and then processes it into rings, necklaces and other jewelry items. Italy has a main jewelry center whose status is comparable to that of Pforzheim in Germany: Arezzo in the south of the Toscana. This is where most Italian jewelry workshops as well as the greatest part of the larger and smaller refineries are located. Italpreziosi is probably the largest Italian refinery and produces its own bars. Another significant refinery worth mentioning is Chimet SPA, but its significance is mainly local.

Gold deliveries usually originate in Switzerland, the nearest home of important refineries. The granules and bars are processed in Italy and then largely exported. Italy's most important distribution market is the United States wherein the quality of Italian-manufactured jewelry is sought.

Unfortunately, the Italian gold market is dying a slow but certain death, or at least its significance will be largely reduced. A comparison of the production volumes achieved in 1998 and 2003 reveals that there has been 60-percent shrinkage. The reason is obvious: high wages and wage-related costs. Italy has been (and still is) losing a large part of its market share to the East, mostly to Vietnam, but also to Turkey. There are obvious parallels to the German jewelry market. Both countries' high production costs in the jewelry industry have undermined their competitiveness.
As its market is exclusively physically motivated, Italy has little significance from a trading perspective. Although the banks, refineries and trading houses involved in the physical delivery take part in international dealings in London or New York they are not major players.

SINGAPORE

The history of the gold market in Singapore started in 1969. After the establishment of the city state in 1965, the government of the time mainly focused on developing into a financial trading center and a venue for the Asian region. Gold trading was handled restrictively, as was the custom in many other countries in those days. Until 1973, private gold imports were illegal in Singapore. Although a tentative trade had begun to develop before then, it only gained significance when—also in 1973—the government lifted its former import levy of three Singapore dollars.

The original gold market had been unregulated and was based on bilateral OTC trades. Singapore nonetheless had great growth potential, as the region is geographically very close to the physical market. Gold-producing Australia is close enough for Singapore to serve as an intermediary or buyer, and Vietnam, Hong Kong, Thailand and Laos represent markets for jewelry producers and buyers that are virtually just around the corner. In the 1970s and 1980s, gold acquisition was a daily subject for central banks and Singapore was positioned at a strategically important location because the central banks of Taiwan and the Philippines, among others, were active in the market as buyers of large volumes. Today, only spot gold and related products are traded in Singapore.

The first attempt to start an official gold futures market occurred in the 70s: the Singapur Gold Exchange. The exchange was established in 1978 and was the first official trading location for gold futures in the Asian region. Its concept was similar to that of CO-MEX or the then still active contracts at the Winnipeg Commodity Exchange or the International Monetary Market. But despite promising conditions, the exchange was not a success. There later followed a gold contract at the Singapore Monetary Exchange (SIMEX, the Gold Exchange merged with the IMM in Chicago, as this syndicate anticipated a greater demand for a gold futures contract), but it was abandoned after a number of years. The contract basically mirrored COMEX's contract: a 100-ounce contract with delivery option and the same purity standards. The greatest problem was that the Asian market has always been and still is very strongly motivated by physical gold. The gold trade is based on a classic jewelry market and has therefore been able to attract a certain speculative potential. But a futures market depends on exactly these speculative market participants to create an effective market.

Singapore has been a purely physical gold center for a long time. As will be explained later on, some trading houses have moved from Sydney to Singapore due to the decline in Australian mine financing and thus added to the importance of the financial center on Malaysia's southern tip in the gold trade.

GOLD

Kilogram bars and Tola Bars
Source: *Christoph Eibl*

There are two types of gold being traded in Singapore today. One is the Loco London, i.e. the classic gold contract, with a valuation date in two days. The other one is the so-called Loco Singapore contract. The quotes are principally based on London Good Delivery bars with a fineness of .9950, but also on the one-kilogram .9999 bar, typical for the Asian region. As the international gold price always refers to Loco London or London Good Delivery bars and a fineness of .9950, the one-kilogram bars are—due to the higher refinery costs, higher consignment and insurance expenditure to Singapore and of course due to the higher purity—more expensive and are traded at an additional premium.

In today's context, Singapore has several clear advantages to offer: It has the long-standing experience as a gold hub for the Asian market. But it is also geographically situated, so favorable that all the neighboring gold-consuming nations may be reached by air within a few hours. Due to the international presence of the bullion banks and refineries, Singapore is able to occupy a central role in the flow of physical gold. It is to be expected that Singapore will not only continue to play this role but could also expand it in future years. Because the largest gold inflow to the Indian market is imported via Dubai and Singapore, and the possibility that trading houses wish to circumvent the region around Iraq, Iran and Pakistan in times of crisis can only strengthen Singapore's potential.

HONG KONG

The Hong Kong gold market has a special kind of supremacy—at least in historical terms. Ever since the Gold & Silver Exchange Society was established there in1918, Hong Kong has been able to proudly boast having the world's very first gold exchange. The Exchange Society originally emerged from the Gold and Silver

Exchange Company. The quality of the gold traded here initially had a purity of .9900. Only later was the fineness .9450 introduced. During British Colonial rule, there were not restrictions regarding the gold trade, except for a few occasions after the Second World War. This was a further great advantage for Hong Kong given the fact that the gold trade has been treated very restrictively during the last century in some countries, sometimes to the point of being completely illegal.

As a British crown colony, Hong Kong has always occupied a central role during large parts of the last century in terms of economic trade activities. It was also very favorably located geographically— just like Singapore. Hong Kong was the portal to the People's Republic of China. Metal and other goods imported to China frequently stopped over at Hong Kong's trading ports and trading houses.

The former Portuguese colony Macao, which borders onto Hong Kong directly, also had an important role to play in the Hong Kong gold trade, the reason being that Macao principally paid a higher price for gold than did Hong Kong. Hong Kong traders hence sold their gold to Macao, from where some of it was smuggled to China, again via Hong Kong. The traders benefited from the price differences between the two colonies. The so-called Hong Kong Macao Gold Trade was very famous because it continued to yield profits for all its participants for almost 27 years. The free lunch came to an abrupt halt in 1974, when Hong Kong's liberalization also opened its market to the international gold trade. The previously mainly local market with its unit of measure Tael and its local gold traders was forthwith augmented by British and American bullion banks, which was more of an impoverishment from some perspectives. Almost all of the important trading establishments, including the Credit Suisse, Scottia Mocatta or NM Rothchild, very soon opened local branches to actively participate in the market action.

The Asian gold market has always been a buyer's market because few ore deposits are exploited in the region, with the notable exception of China. The produce of South African, Australian or North American mines could hence usually find a buyer here. And because most British and American trading houses were closely linked to the mining industry, there was a great interest in a closer collaboration with the physical distribution market in Asia.

GOLD

In the wake of the establishment of the gold exchanges a local trading and market association was founded—comparable to London's LBMA—to ensure a representation of interests and determine market standards: the Hong Kong Bullion Dealer Club (HKBDC). This association mainly served to strengthen the relationships between the major trading centers. European and American traders and bank representatives regularly visited Hong Kong to put further emphasis on the mutual exchange of interests and knowledge.

Because of its Chinese origin, the Hong Kong gold market differs from other trading centers, at least as far as its unit of measure is concerned. This means that Chinese units of measure are used. The unit of measure Tael equals 1.20337 troy ounces. In Germany, the most popular bar is the one-kilogram bar with a fineness of .9999, whereas in the Chinese region it is the 20 Tael bar with a fineness of .9900.

Because the Hong Kong Gold Exchange has radically changed in recent years, I shall examine this market and its structures in greater detail.

Similar to Singapore, there are basically two types of gold traded: Loco London and Loco Hong Kong. Loco London in Hong Kong is the same as in the London gold market. The benefit usually lies in the time difference. Traders and gold mines can, for instance, place overnight orders in this market that will benefit from price fluctuations on the Asian markets without requiring an Asian infrastructure (vaults, insurances, agencies). Loco Hong Kong gold functions differently and conforms to the characteristics of the local market (differences in valuation, trade volume and execution). But Loco Hong Kong gold is also being traded in the trade centers of Singapore and Indochina.

One of the most important differences is the Saturday trade. Whereas all the other trade centers around the globe are closed over the weekend, Hong Kong's exchange stays open on Saturdays, albeit for a shorter period of time. Not that many market participants seem to be fully aware of this fact.

Hong Kong is also the only gold market featuring same day settlement for all physical gold trades. Just for comparison: The classic gold trade is usually performed with spot contracts, i.e., with a settlement period of two days. Although same day settlement is usually available in Hong Kong, quite a few trades are only settled some days later. A special interest formula is used to calculate the holding expenditure. This formula is fixed daily by the exchange and published in Hong Kong Dollar per 10 Tael. This interest is called the Carried over Charge (COC) and not only includes the usual financing costs (usually via the Hong Kong Dollar interest), but also a premium according to the supply and demand situation. If there are market shortages, the premium rises accordingly and falls again, as soon as enough gold is available once more.

Trading Hall of the Hong Kong Gold Exchange
Source: *The Hong Kong Gold Market*

The COC is also divided into five classes: The first class of COC applies if there is a shortage of Tael bars on the trading day and the seller prefers a deferred settlement. If the shortage still persists on the fourth day, the second calls of COC will apply from this day onwards. This procedure is continually repeated until the fifth class is reached. The increases associated with each class are fixed at approximately 4 percent, whereas the initial interest is approximately

10 percent. The fifth class is only rarely reached because the costs of the COC would then be too high to justify a longer period of deferral.

The purpose of these interest rates is to ensure the possibility of physical delivery. A kind of penal interest rate forces the sellers to deliver as soon as possible.

In addition to the continuous trading activity there are also two daily fixings. (On Saturday, there is only one: the morning fixing at 10:30 a.m.) The morning fixing starts at 11:30 a.m. and the afternoon fixing starts at 16:00 p.m. local time. The fixing procedure is similar to that employed in London, only that there is no flag process (see the description of the fixing in the London market) and that fixing interest described above is negotiated in case of an imbalance of fixing interests. The fixing interest applies to all open positions in the market.

This demonstrates another significant difference to other cash markets. The cash market is based on a margin system, not unlike that of a futures market. Many market participants hence do not speak of gold purchasing per se but of contracts referring to gold acquisitions. Just like in the forward markets in Japan or New York, the Loco London contract is based on an initial margin and a maintenance margin, as the following example illustrates:

The trader who buys a contract (100 Tael) has to deposit the respective initial margin, and the gold is posted to his account. Depending on price fluctuations, the trader receives a margin call if the value of the long position has dropped. The interest applicable to the position in his account (similar to a drawing account) is added to or subtracted from this account every day.

Experience has shown the trade to be quite different from the London cash trade. Because there is no need to pay in the whole sum for the gold purchase, market participants are able to leverage their trade volume a lot more. In some cases this leads to relatively high turnovers. But the trade is always beset by a large number of execution and monitoring units. A thorough knowledge of the local market is of paramount importance. It is, for instance, not only important to know which bars may be delivered for the contracts, but also which securities or bonds may be paid in against margin accounts.

After the Shanghai Gold Exchange opened in 2002, many market participants became convinced that China was trying to control the gold trade from the People's Republic or diminish the dominant position of the Hong Kong market. The developments of recent years seem to prove them right, as the Shanghai market is currently growing very fast indeed.

CHINA

October 20, 2002, marked the start of the first free—at least for Chinese standards—gold market in China. On this day the Shanghai Gold Exchange (SGE) opened its trading floors. But why should this event be so important? Hadn't countless stock exchanges tried in previous years to trade in gold or at least list gold contracts, often failing pitifully?

To answer these questions one needs to look back to the past. At the start of the 20th century, China had not been a major player in gold production—silver was still the primary instrument of trade—which also explains why the Chinese central bank still has such large silver holdings today. And as the sway of the Chinese communist regime triggered massive inflation, the authorities simply outlawed gold ownership. Anyone still found in possession of gold faced stiff penalties. All private gold holdings were to be exchanged for paper money. Ever since then, the markets continued to be controlled by restrictions, which intermittently managed to push the local Chinese gold price $50 USD above international price levels. The price spread had been caused by import restrictions, duties, taxes and compulsory levies.

Then, in the 1990s, the World Gold Council, the gold mining industry's lobbying organization, finally began making itself heard in China. Talks were held with representatives of the People's Bank of China and other public officials. The first breakthroughs were achieved in the middle of the 90s when the gold tax was halved. In the end, the Chinese authorities relinquished their control of the gold price completely in 2001.

The gold exchange started in 2001 with a test phase, so to speak, to see whether legalization could stem the flood of smuggled gold.

GOLD

This endeavor was particularly important because Hong Kong had always been a major hubcap of the gold trade, also for China. Yet all market participants agreed in their disdain for gold smuggling because the clash of fixed and free market prices invariably creates arbitrage opportunities, which, in this case, could only be exploited illegally.

The premiums on gold were significant in China compared to the Hong Kong market or the free international gold market. With a number of refineries at its disposal, Hong Kong represented the Far East's primary trading center. And the high volumes of the gold trade meant that gold could be smuggled into Vietnam, the former Portuguese colony Macao, and also into China. It was these smuggling operations that both Hong Kong and China wished to stop. China therefore started to establish its very own institutionalized market with international support led by banks from LBMA circles.

The opening of the gold exchange thus marked the Chinese government's incremental withdrawal from the distribution process of physical gold. One of the greatest advantages institutionalized trade has to offer is the determination of an internationally reproducible price, thereby creating greater price transparency and acceptance. This also took effect in China, where the local gold price's development, after only one year of free trading, closely mirrored that of the international financial community. Extremely high premiums of more than $30 USD disappeared altogether, whereas marginal premiums and discounts may still be observed in the market even now (last year, prices ranged from $12 USD above par to $9 USD below par), and the smuggling has been on the wane for several years thanks to the free trading opportunities. It won't disappear altogether, though, as some price differences still occur.

The Shanghai Gold Exchange offers a cash market dealing in the finenesses .9999 and .9995. The gold is traded in Renminbi per gram and kilogram. Vendible volumes are the one-kilogram, the 100-gram, the 50-gram and, since 2003, the 12.5-kilogram (400 ounces) bar. There is also a three kilogram bar which is meant to represent the equivalent to New York's 100 ounces contract (3 kilograms equal 96.31 ounces).

Trading Hall of the Shanghai Gold Exchange
Source: *The Alchemist*

China has two refineries licensed by the LBMA. These may produce gold bars that could theoretically enter the global gold trade without much of a problem. In contrast to the global trade, the Chinese gold trade is a (t+0) trade where contract placement and delivery occur on the same day. A deferred payment method (t+5) is in use, as well. Gold may be delivered to or collected from 12 official depots. Only gold traded via the exchange is VAT exempt. This measure is, among other things, intended to attract new investors.

There are currently no options in gold, although the exchange is well-equipped for a complete futures trade. Futures trading has been introduced briefly but was abandoned very quickly.

The World Gold Council pursues the stated aim of using determined lobbying to create a completely free gold market, which would easily be absorbed into the international gold market network. But this would also mean that the country's currency, the Renminbi, would have to be freed from trading restrictions. China's international importance should not be underestimated: It is the world's fourth largest gold producer and now experiences an enormous demand for gold jewelry, for instance, which can only intensify in the future given the strong economic growth.

In 2004, the LBMA symbolically underscored this fact. In order to draw attention to the international significance of the country's gold market, the event committee responsible for planning the world's largest annual bullion conference decided to hold the very first gold conference in China. The decision was met with great applause, including from the Chinese side.

AUSTRALIA

Gold trading starts in Australia—at least in chronological terms. Every morning's first gold quotes come from Sydney. But although the market determines its first gold price of the day here, turnovers are not very large and also limited to the local market. There is no comparison to London trading time.

Australia is a classic gold mining country. With its 300 tons of gold produced annually, Down Under ranks third globally. The large open pit mines are characteristic for Western Australia, the region around Kalgoorlie and Perth. The gold is in Australia much nearer the earth's surface than, let's say, in South Africa. It is usually found at a depth of 200 meters, whereas South African gold mines frequently need to drill several kilometers deep to get to their gold ore. One Australian specialty is the so-called nugget fields, where the gold is not combined with other ores but occurs loosely. At least in the past this meant that the gold could be mined relatively quickly and easily.

Open Pit Mine
Source: *Christoph Eibl (Photo)*

Where there is gold mining, the banks are usually not far off. They serve to provide capital, advice and sometimes act as trading partners for gold trades. Because mining development and exploration is very cost-intensive, the classical mine financing is needed as much now as it ever was. To receive finance, mine operators are required to provide so-called banking feasibility studies. Their main purpose is to prove that the anticipated gold is actually there and can be exploited economically. The banks usually demand that a part of the anticipated gold production is sold up front as a forward trade to ensure that the mines will be able to fulfill their credit obligations. The process is also called credit hedging. In this case, the hedging serves to ensure that the business plan, which is based on anticipated gold prices or, respectively, calculated on the basis of current gold prices, can be realized, because the gold price can be subject to relatively significant market fluctuations.

Banks could benefit from these deals in more ways than one (and they certainly can still do so today, even if the hedging demand in project financing has strongly waned in recent years):
First, from the loan and the margin resulting from it; second, from the hedging deal itself. An example:

A mine requires $200 million USD to build a mine complex. The mine is expected to produce a total of 800,000 ounces of gold over the following five years. The gold price is $450 USD per ounce. The bank willing to finance such a project now calculates several scenarios. At a constant gold price of $450 USD, the project would yield a gross amount of $360 million USD (minus the $200 million USD investment, not including current expenses and interest charges) before cash cost, i.e. the mining cost. To secure the loan with collateral, the bank will ask the mine to fix the gold price to $450 USD for parts of its production. That can happen in a variety of ways. They could, for instance, sell 400,000 ounces right away, which would fix the price. Because the mining company does not currently have this gold, it needs to borrow it. This is where the bullion bank comes into play. The bullion bank finances the credit via gold lending. As soon as the production is underway, the mine reimburses the gold to the bank. This kind of trade can be optimized via forward acquisitions or sales and also via options. The illustration below shows the procedure of gold production and illustrates which parts of the process the banks are involved in.

GOLD

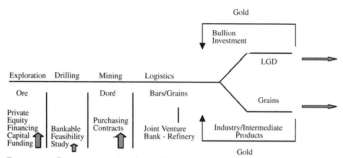

Process Sequence in Gold Production
Process Sequence of Gold Production
Source: *author's own representation*

The Australian gold trade is therefore highly dependant on mines and bullion banks, and their various engagements. The strongest players in the Australian market are the Australian banks Macquarie and the Commonwealth Bank of Australia. There are, of course, several brokers at hand in Sydney providing the market with the required liquidity.

Globally positioned banks usually maintain a trading branch in Sydney, Singapore or Hong Kong to cover the Asian-Australian region. But in the last few years, a number of banks have withdrawn from the Australian market. In particular, the refocusing of some banks on their core business has been to the detriment of mine financing in some financial institutions, as the mine business is nowadays of central importance in very few banks. In addition to this, the increasing importance mining companies attach to shareholder value has made the subject of hedging a thing of the past for mine investors.

JAPAN

Japan has little to no local gold deposits. The trading establishment Sumitomo is a shareholder in Japan's only gold mine, which produces a few tons annually. The Japanese jewelry and electronics markets are hence forced to act as gold buyers and importers on the international market. Australia used to be the main supplier in the past, delivering either directly to Japan or via the two large trading centers in Singapore and Hong Kong.

Over a long period of time, the Japanese gold trade used to be strictly controlled by the Bank of Japan. All imports were initially processed by the central authority, the Japanese ministry of finance. The ministry very effectively exploited this monopoly by buying gold at international prices and then selling it in Japan at a high margin. This type of arbitrage, which sometimes raked in profits of over 50 percent, was only stopped in the beginning of the 1980s when banks and trading houses were finally given permission to trade and import gold themselves.

This also directly resulted in the establishment of the Tokyo Gold Exchange in 1982, which is now called the Tokyo Commodity Exchange (TOCOM) after a merger in 1984. Although the TOCOM was repeatedly forced to adjust its contracts in its early years, it is now, together with COMEX, one of the most important forward exchanges for the trade in precious metals.

Trading Hall of the Tokyo Commodity Exchange
Source: *Tokyo Commodity Exchange*

In addition to gold, the metals platinum, palladium and silver are also traded at TOCOM, with the largest turnovers occurring in the platinum trade today. Gold is traded at a fineness of .9999 and quoted in grams. The contract is based on grams as the unit of measure and on the Japanese Yen. The current seven straight months are traded here, which means that in March 2005, for instance, April 2005, June 2005, etc., are traded. In contrast to COMEX, where the nearest contract is always the one to record the largest turnover, the contract that is the furthest off chronologically is most actively traded at TOCOM, reason being that Japanese speculators assume that they have more time with a longer-term contract than with a shorter-term one.

In the beginning there was no permanent trade, but six daily fixings instead. But in 1991, the exchange introduced an electronic trading platform that is geared for continuous trading. Here, however, are still some differences to COMEX. The most important is probably the "Limit Up" and "Limit Down," which means that price fluctuations may not and cannot exceed or undercut a certain price level within the same trading day. As soon as a Limit Down has been reached, prices cannot decline any further on that same trading day. Anyone still interested in trading will have to make do with this Limit Down price. The same applies to Limit Up in the opposite direction. Trading starts at 8:20 a.m. Tokyo time and ends—after a break from 11:00 a.m. to 1:00 p.m.—at 3:30 p.m. But brief fixings are held for each contract term before the start of each trading segment. There is also a pre-order phase that starts half an hour before the respective trading segment, during which orders may be placed.

The market participants are grouped in three distinct categories: speculators (the general public), arbitrage traders (trading houses and overseas bullion traders) and locals (future commission merchants, brokers). These supply the required liquidity in the various contracts and maturity terms.

Similar to COMEX, fully-fledged trade participants require a seat at the exchange. In the early 1990s the cost for these representations at the exchange were so prohibitive that virtually only local trade participants were able to afford one. Membership requirements have remained very strict to this day in order to ensure that only the best may trade in this location. The largest local players are Mitsui & Co., Sumitomo and Mitsubishi. These trading establishments have by now established offices and trading units all over the globe. Japan also has several refineries, with Tanaka K. K. probably being the most important.

In August 2004 the Central Japan Commodity Exchange (C-COM) announced the start of a gold futures contract denominated in USD for 2005. This will be the only USD-based future in the entire Asian region and is guaranteed to create a huge demand in the international trading community. The contract is intended to be based on ounces and settled via cash settlement, as opposed to physical delivery. The application for permission to trade this contract is currently under consideration at the Japanese Department of Commerce. The C-COM is clearly trying to tackle the issues associated with the Yen/gold contract quoted at TOCOM by changing the trading units and currency: Yen to Dollar, and grams to ounces. There have been repeated attempts by various exchanges around the world to bring liquidity to gold futures. Many of them failed—but Japan would henceforth enjoy two contracts in gold. It remains to be seen which of the two will turn out to be the winner. In conclusion, it is worth mentioning that the TOCOM gold contract is not the most important, despite its monetary supremacy (because much higher turnovers are achieved, predominantly in platinum), but that the global trade's first exchange price is determined here. Japan is a typical import (i.e. gold buying) country and will hence always play a significant role in the global gold trade.

THE FORMER FRENCH-INDOCHINA

The former French-Indochina is today comprised of Vietnam, Thailand, Laos and Cambodia. During a large part of the 20th century, the region was plagued by the French/Indochina War, then by the American/Vietnam War and by the horrors perpetrated in the name of the Khmer Rouge regime in Cambodia. But the region always remained, and still is, very important to the gold trade—particlarly because Vietnamese jewelers are world-famous for their creativity. In the 1970s, Saigon (now Ho Chi Minh City) became a world leader in terms of innovative designs and high product quality. During these difficult times, gold was also used as an alternative currency in Cambodia and South Vietnam and was the preferred mode of payment for most "normal" traders. One good example for this is the 0.5 centimeter-thick Kim Than gold bar that was in use as a local currency.

Because of the communist or, respectively, totalitarian regimes, a free gold trade had not been possible in Vietnam for many years. The gold market was finally deregulated in 1999. The State Bank of Vietnam (SBV), which was responsible for the gold trade, had up till then always provided local traders with bid and ask quotes for gold. It was very difficult for the gold traders in those days to effectively and accurately balance gold volumes, as the quotes were often apportioned without rhyme or reason. The impulse to government deregulation finally came from the World Gold Council, which opened a Vietnam branch in the 1990s.

Since then, Vietnam has developed into one of the most important importers. The country is of greatest interest to international jewelry manufacturers due to its low overheads. And Vietnam is furthermore located directly on the doorsteps of Hong Kong, Shanghai and Singapore. The government finally authorized some state banks to trade in gold options in December 2004. This decision is undoubtedly a great boost for the Vietnamese market. But on the down side, the country's market is still subject to tentative regulations by the State Bank of Vietnam. Gold that is to be processed, i.e., gold that is imported to Vietnam and transformed into jewelry there, is required to be exported again after this value is added. That is the law. This might enhance the country's employment situation, but it is an obstacle for many market participants. The country's efforts to become an internationally significant gold buyer are clearly recognizable. In the international trade, Vietnam is currently primarily significant as a gold importer. Globally positioned bullion banks maintain local custodian warehouses or distribute their gold locally via their refineries.

The main brunt of these Vietnamese deregulations was probably borne by the Italian market. Some of the gold and silver that was previously transformed into jewelry in the Italian gold province Arezzo now goes to Vietnam. The main reasons are obviously the attractive employment terms and the excellent know-how of Vietnamese jewelers. It is much cheaper nowadays to produce gold jewelry in Vietnam than in the German Pforzheim or the Italian Arezzo.

All in all, Vietnam and its neighboring countries today play a similar role to that of India, even if gold ownership is, in terms of tradition, less important in Indochina than in India.

RUSSIAN FEDERATION

As a trading location, Russia was of minor importance in the past. The former Soviet Union relied on other markets and frequently only appeared as a gold distributor. In contrast to many other countries, Russia is not only rich in raw materials, but particularly rich in gold deposits. Even in the 1980s and 1990s, the Soviet government sold reserves or newly mined gold on the international market, frequently via Switzerland, in case of harvest failures or financial bottlenecks.

Up to the 1990s, all gold producers from the Federation of Independent States (GUS) were forced to sell their gold to government authorities, which then sold it in the market or developed their own state positions. The gold producers were either under close state observation or run by the state itself, so that a free market was as unthinkable as a free trade itself.

But the government changed its tack in the beginning of the 21st century and granted so-called gold export licenses to certain banks. The banks that had been thus granted licenses bought up the gold from the mines and sold it again in the international market. They also gained access to products such as options or forwards and were put in a position to pass these on to the domestic industry. The central market regulation was hence incrementally withdrawn.

Today, Russia is the fifth largest mine producer in the world. Market participants are convinced that Russia and its trading location, Moscow, will be of significantly increased importance for global trading in the future. The opening up of the market to foreign gold companies puts the Russian gold market in a position to rapidly discover and exploit gold deposits using sophisticated foreign methods and know-how. The latest acquisition of 20 percent of the fourth-largest South African gold mining company, Gold Fields, by the Russian metal conglomerate Norilsk Nickel shows how interested Russian market participants are in expansion and the acquisition of mining knowledge.

The liberalization of the market is also useful on a trading level. Many deals, that until recently had to go via Zurich or London, are now—after some infrastructure development—also thinkable in Moscow. The national pride of many Russian banks naturally tempts them to draw as much knowledge as possible from the international market and use it on their home turf. Until now, the major Russian banks, such as Rosbank and Sberbank, were active in the gold market. They financed local gold projects and carried out gold sales in the international market.

It is quite predictable what will happen to the market with Western capital and know-how: increased knowledge and the attention of international companies will strongly support not only the mining sector, but also the gold trading sector, thereby augmenting Russia's importance as a whole. That one shouldn't underestimate the Russian market is clearly demonstrated by the LBMA who held a gold conference in Moscow last year.

SOUTH KOREA

Our detour through the international gold markets ends in South Korea. Although the gold trade is of secondary importance in the country, Korea has had a functioning gold forward market for years, the Korea Futures Exchange. The Gold Fields Mineral Service (GFMS) describes the South Korean gold market as one of the most interesting in 2003, as the value of the gold jewelry exports, albeit based on a smaller volume, rose by over 2,000 percent to $630 million USD between 2002 and 2003.

The Korea Futures Exchange has offered a gold contract since 1999. The contract traded is denominated in one kilogram and a fineness of .9999. It is being traded in the Korean Won. Anyone trying to find arbitrage potential will have to compare the Korea Futures exchange to the TOCOM or the Hong Kong gold market, because these exchanges are situated in roughly the same time zone. But arbitrages will probably fail in practical terms, as there are three currencies to be considered, in addition to the rather low liquidity. But it is still interesting to see that the contract has held its ground for a number of years, despite the Korea Future Exchange's relative isolation from international markets. In global competition it will always be relegated to a secondary role, though.

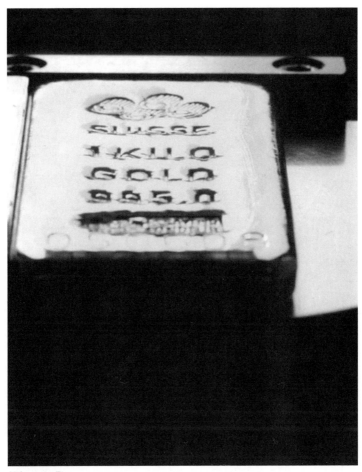

Kilogram Bar
Source: *PAMP SA*

The Most Important Trading

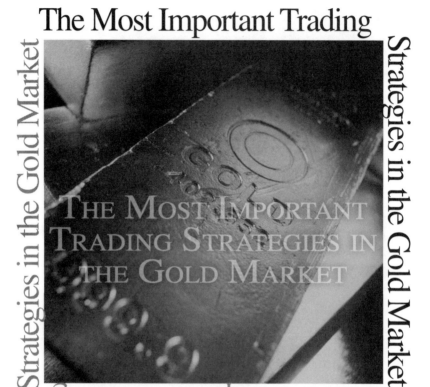

THE MOST IMPORTANT TRADING STRATEGIES IN THE GOLD MARKET

- III -

Precious metals markets do not work the same way as other financial markets. There are two reasons for this. First, different market rules apply to gold than to shares or bonds; second, the market relationships valid in the international trade for gold are quite different from those applying to the remaining precious metals, which was clearly demonstrated in the country-specific sections. As there is a whole variety of products that may be used in the gold market, it can be possible to pursue futures strategies, cash strategies or a combination of the two.

At least theoretically, the possibilities are endless. In efficient markets, in the sense of capital market efficiency after E. Fama, one could assume that arbitrage is practiced wherever it is possible.

This doesn't mean that the gold market is some kind of free lunch for arbitrage traders, but the market differences are usually equalized in very short time after appearing. One must therefore assume that arbitrage traders are active in the markets, observing exactly these market imbalances and exploiting them, given the chance.

The precious metals market—and this will more clearly emerge with the broad range of arbitrage options—is not as efficient as the bond market. Whereas classical markets rarely offer opportunities for arbitrage trades—and if they do, they're mostly highly complex—one can be quite successful in the gold market with a relatively simple trading strategy, such as cash-and-carry arbitrage. I shall explain some of the most common trading strategies below, but let me first start with a theoretical explanation, before the individual strategies are presented, using practically relevant figures. I will only demonstrate a fraction of the various trades practiced in the market. This excerpt is by no means exhaustive; rather, it explains various products, how they are used, how cash and futures markets are interconnected, and how the physical component makes these markets so unique.

Let me first of all explain the term arbitrage. Arbitrage is the attempt to make a risk-free profit by exploiting differences in prices quoted for an equivalent good at two different exchanges. To this end, the good—in our case the gold—needs to be homogenous in itself. This condition is certainly fulfilled, but historically grown structures and the susceptibility of gold to reshaping can somewhat complicate the exchange.

As we have already seen, gold is traded globally, but the trade is based on various finenesses, sizes, currencies and units of measure. It is therefore necessary to formulate a common reference value in order to explore which type of arbitrage trade might be possible. And because gold is traded in USD and ounces during two thirds of the available 24 hours, these values shall serve as a basis for the following. The cash price hence always needs to be adjusted to one ounce and USD. The conversion costs this causes need to be included in our comparative calculations. They comprise the spreads applying to currency conversions, the adjustment of logistical depots, bar sizes and other conditions still to be explained.

Pure arbitrage is most commonly examined in reference to gold and its price, but often the more interesting profits are found in factors like foundry and consignment costs or interest rate conversions. More on this will follow later on in the chapter.

The most important thing is to find a common price source in order to check the strategies' consistency. An ideal candidate would be the information and news service Reuters with its sources for currency rates, interest rates and gold prices, whereas consignment, insurance and storage costs will have to be queried from direct tenders. Brinks, ViaMat and MSK SA are extremely reliable, also in terms of quoting prices and performing the relevant transactions.

It is always on this cost background that individual gold market trading strategies are analyzed.

ARBITRAGE TRADES

Currency Conversion Costs

Currency pairs are always supplied with a buying and a selling price (double-barreled). These depend on the market situation, order size and, of course, on the currencies in question. While, for instance, the Euro can be sold against the USD at a spread of 0,0001 Dollar, this kind of low spread can hardly be represented in conversions of Indian Rupees into Dollars. The currencies relevant to the arbitrage therefore need to be checked for costs referring to the spread.

Theoretically, a currency risk always occurs whenever gold is invoiced in another currency. One should—again, theoretically—hedge the position against this risk via forward contracts or options. But these currency positions in gold are usually only held for brief periods and can therefore generally be discounted. This is, of course, only valid for short term oriented trading strategies because these risks can gain added relevance in forwards or longer term trades, such as spread trades, and should therefore be mentioned here.

Location Swap Costs

Location Swaps simplify metal consignments and generally reduce costs. Let us assume that someone has a gold account in London and would now like to send the gold to India to be able to access it locally in Mumbai, for example. The resulting cost calculation would need to include consignment from the vault to the airport, airport levies, flight costs and consignment costs, to the final location in India. These expenditures can be apportioned to the goods to be transported, resulting in a certain amount per ounce. This rate is the basis of quotations in location swaps.

This is a purely mathematical representation of the situation. In reality, consignments can work out to be much less cost-intensive because some trading establishments maintain gold depots in both locations and are able to perform a purely electronic posting. The price for this kind of swap can hence be something like $0.25 USD per ounce, for instance, assuming that it concerns a total volume of 500 kilograms.

But how does the physical location swap actually work? The international gold market principally trades Loco London. The mines, which usually produce their gold in much further-flung regions and wish to sell it, are forced to transport it to London. Some trading houses buy the gold directly in situ, for instance, in South Africa, at a discount representing the consignment cost from South Africa to London. The actual consignment then sometimes goes directly to the end consumer, in this case to India, for instance (insofar as the gold is already in the appropriate end sale state). The gold hence follows a direct path, with costs which can usually be more than covered by the discount to London. Because mining companies and trading houses frequently maintain long-term delivery or buying contracts, traders continually have stocks in India, which they can, of course, balance with gold positions in London. The location swap functions on this background as follows: A bank wishes to sell its gold, which is currently in London, to a jewelry manufacturer in India. As the physical transport costs would be prohibitive in terms of time and money, the trading house that has executed the South Africa-India trade mentioned above would offer its services as a swap dealer. The mathematical cost will then be virtually the same, but at least one wouldn't have to bother with customs and tax questions. The trading house would offer the bank the gold in

London, and as a compensation credit, the same position again in Mumbai, India, for a price of $0.25 USD per ounce, for example. The whole procedure would be done electronically, i.e., the gold would not be restacked and transported.

Just as an indicator, the following prices for location swaps can be considered:

From	To	Cost per Ounce	Quality	Bar Type
London	Zürich	0,15 US-Dollar (varies strongly)	.9950	LGD
Zürich	Mumbai (ex Bombay)	0,30 USD	.9950	LGD
Johannes-burg	Mumbai (ex Bombay)	0,25 USD	.9950	LGD
London	Istanbul	0,25 USD	.9950	LGD
London	Dubai	0,25 USD	.9950	LGD
London	Singapur	0,25 USD	.9950	LGD
London	Hongkong	0,30 USD	.9950	LGD
London	New York	0,50 USD	.9950	LGD
Dubai	Mumbai (ex Bombay)	0,30 USD	.9950	LGD

Prices of Global Location Swaps
Source: *MKS SA, Brinks, author's own research*

The table refers to deliveries of 500 kilograms (16,000 ounces) each. For the deliverer, CIF applies (Cost Insurance Freight, pursuant to the international INCO Terms, which are defined by the International Chamber of Commerce) from/to airport warehouse.

Quality Swap Costs

These concern cases where the unit of weight needs to be changed by means of cost-intensive recasting. For instance, if gold with a fineness of .9950 is to be turned into fine gold with a fineness of .9999, one speaks of a quality swap. If no tender—who might have various qualities of gold in stock—can be found, the swap may only be carried out by smelting. One can generally say that all geographical locations exact their own specific requirements concerning gold, which also means that there are a sufficient number of refineries producing .9999 fine gold as well as gold of lesser purities. One basic rule applies: If there is supply in the market, quality swaps may be performed without significant costs (except for the higher gold value). But a lack of supply creates costs depending on the gold's volume and original quality. In this case, the time factor is also to be considered, not only in terms of labor time but also regarding refinery capacities.

Size Swap Costs

This type of swap concerns the adjustment of bar types and units. The gold price fixed in the London market within the LBMA is principally based on 400-ounce bars. Their minimum purity is .9950, whereas the Indian jewelry market, for instance, requires smaller sizes, the so-called T.T. bars (10 Tola bars) or, increasingly, one-kilogram bars. Adjusting bar sizes and units of weight can therefore become necessary to satisfy specific client requirements. There is usually a supply of bars of all sizes and qualities in the market. Bottlenecks generally only occur in certain periods of increased demand. One can principally say that the costs created by this type of swap tend to be insignificant per bar and unit. In a sample transaction executed in December the following prices were determined:

Tael Bars
Source: *Rand Refinery*

103

Swap of 10 London Good Delivery bars Loco London for one-kilogram bars (.9950): The refinery in this case invoices a price that only refers to processing costs, because it can keep the foundry's furnaces busy with the London Good Delivery bars and turn them into other bars. The swap therefore cost $350 USD ($0.0875 USD) per ounce (10 LGD bars are the approximate equivalent of 4,000 ounces).

Storage and Insurance Costs

In London, the gold hub, costs are incurred for the storage and insurance of gold placed in allocated, as well as unallocated accounts. Apart from the Bank of England, which safeguards the gold of many nations (most of them former colonies or current Commonwealth members), there are five further general clearers offering gold and silver management. Although prices tend to vary they stay within a relatively narrow range. The HSBC London, for instance, currently charges 0.05 percent per annum for the storage of unallocated gold. This includes the insurance. The allocated storage of gold tends to be significantly more expensive and usually costs between 0.30 and 0.50 percent a year on the inter-bank market. This is also the reason why allocated storage is only to be recommended if there are good reasons or it is specifically desired. The costs result from the separation of the gold from all other assets, which is very labor intensive. The bars are appropriated, their numbers are communicated to the owner, and then they are finally stored in a vault, physically separated from other bars.

This is required knowledge for anyone wishing to trade efficiently in the gold market. The general character and in particular the special traits of gold as a raw material or good were highlighted.

A screening of the inter-bank market or retail departments would reveal that there are some market participants who pursue pure trading strategies and, on the other hand, the so-called arbitrage traders. Arbitrage is basically a trading strategy, but it is based on the idea of benefiting from price spreads in the market without incurring any risks, whereas trading strategies frequently also include open risk positions.

Let me give you an example: A trader compares the prices quoted for one and the same product in two different exchanges or market locations. If he detects that the gold is cheaper at trading location A than at trading location B, he will try to buy the cheaper gold and sell it again in the other trading location. The whole trade is highly dependent on logistics, as we have seen with the components Loco swaps and consignment. If the profit earned in our example exceeds the costs incurred through logistics, for instance, the trader would be able to make a profit without incurring any risk—at least in price terms—as operational risks can almost not be excluded in this case.

But as this simple constellation is obviously familiar to all market players, this type of profit is rarely achieved.

In reality, trades of this type are further complicated by currency and weight components. If you compare the prices quoted in Istanbul and London, for example, the currency component would be Lira to USD and the weight component ounces to kilograms. And that still leaves the question of consignment from Istanbul to London to be considered.

Only if you are able to break down all the prices traded globally into a single standard unit can you get a reasonably good idea of how markets trade gold at a surcharge or a discount.

Costs Created by Operational Risks

Operational risks do not create costs in the literal sense, but definitely need to be included at this point as they can occasion direct costs, should worse come to worst. Operational risks are risks incurred through the very execution of the trade. They include the risks of contracting party failure and late delivery. In these cases, interest penalties and value adjustments would be the result. But to include all of them in the calculation for each arbitrage trade would be going too far and probably eliminate any profit. They are only listed at this point for the sake of completeness:

- Transaction risk (shipping, customs or deposit);
- Risk of contract party failure (the party with whom a binding contract was made does not fulfill compliance obligations);
- Execution risk (the wrong prices are confirmed or invoiced);

- <u>Delivery risk</u> (gold is delivered too late, resulting in interest penalties); and
- Quality risk (gold is delivered in inferior condition or could even be forged, for instance gold with a Wolfram core).

Loco Arbitrage in the Cash Market

As the 24-hour table below shows, only certain locations can be arbitrated simultaneously in Loco arbitrage, as not all time zones overlap. Simultaneous trading is only possible for a few minutes a day between the Japanese and British markets, for instance, whereas in the Asian time zones, several trading locations are active simultaneously, which means that it is relatively easy here to calculate price spreads.

To this end, the following table should be consulted first to make the markets a little more transparent.

	TOCOM	CGSE	SGX	loco SG	loco SY	IGX	LBMA
Currency	JPY	HKD	CNY	US-Dollar	AUD	TRL	US-Dollar
Unit	KG	Tael	Tael	Ounce	Ounce	KG	Ounce
Contract	1000 g	10 Tael	10 Tael	#	#	1 kg	#
Division of	/32,105	/1,20337	/1,20337			/32,105	
Price per ounce	*JPY/USD	*HKD/USD	*CNY/USD	#	*AUD/USD	*TRL/USD	#

Gold Exchanges and their Conversion
Source: *Author's own research and illustration*

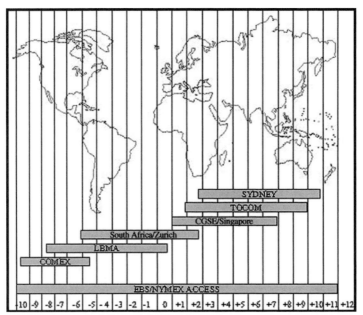

Gold Exchanges and their Global Trading Times
Source: *Author's own research and illustration*

Let's take an example dated November 25, 2004:

Tokyo one ounce (converted) $444.30 - $444.80 USD
Hong Kong one ounce (converted) $443.80 - $444.10 USD

It was hence possible at 12:15 p.m. (Tokyo time) to buy gold at a cheaper price in Hong Kong than in Tokyo and simultaneously sell it in Japan at a profit of $0.20 USD. As long as this type of trade is executed at exactly the same time, one can initially perform risk-free arbitrage of 0.20 USD, minus the spread. The above mentioned costs of quality swaps, size swaps and Loco swaps are only incurred if the physical delivery of both contracts is included. The first thing to consider is that bars need to feature the relevant assayer's (i.e. refinery's) grade. Because at TOCOM in Japan, only specific bars from authorized refineries may be committed to gold contracts. One also needs to consider for this trade that the TOCOM is a futures market and that the spot price was hence rated. Like in any futures market, the rating is calculated by extracting the lending yields or costs, respectively, and the financing interest (more on this in chapter 5.4 Cash and Carry Arbitrage).

GOLD

In purely mathematical terms, not counting the consequential costs, a profit is made. But as soon as these costs are included one discovers that the spread of the gold price in itself has not come about accidentally, but for a very good reason: the extra costs created by swaps. This is because the prices are not only affected by the refinery's recasting and refining costs, but also by the following components:

- Local availability of bars/shortage/excess;
- Extent of total transaction volume (higher volumes lower unit costs); and
- Capacities of logistics and security companies.

Due to the geographical distance between the exchanges, arbitrage deals are beset by timing and logistics problems. If the trade is between London and Istanbul, for instance, one has to usually consider a time period of two days from contracting to fulfillment of the deal. This is called spot valuation, which was explained above. If the gold was bought in Turkey and sold in London, the trader would receive a consignment of Loco Istanbul on the second day and simultaneously owe a Loco London delivery on the same day. This can only be achieved logistically within the relevant time at a very high cost. The trader therefore needs to simultaneously arrange for the Loco swap on the day he enters the trade. He is hence looking for a swap trader whom he can supply with the gold in Istanbul and who will receive it in London. The problem of geographical distances, and therefore of the time period to be included in the planning, is even more significant between London and Hong Kong, for example.

The costs incurred by the actual transport and physical recasting usually eat up the price profit achieved. Therefore, another arbitration method needs to be found. One variant consists of maintaining various depots at relevant market locations. The basis is usually an account with a general clearer in London. Depending on the focus, an account in Istanbul or Hong Kong might be desirable. Now the electronic posting of gold trades can be represented more easily and stocks can be simply transferred via electronic swap. The result: Costs are lower and trading strategies may be realized more efficiently. Though, one factor to consider in this scenario is the storage costs at the depots. But depending on the volume of the trades, this effort in terms of cost and labor may be worthwhile.

Arbitrage trades can also, of course, be executed with just one depot, for instance in London. All the other locations will then be delivered via Loco swap. But it's evident that costs are highest in this case, which means that the theoretically achievable profit is only rarely realizable practically. Another solution would be to exclusively trade with electronic products without ever getting involved in the physical side of the business. Futures trades would be the ideal candidates for this. But on the other hand, this type of trading completely excludes numerous markets and trade volumes.

One of the most important questions in the world of arbitrage is: "What is the price and is it fair value?" The gold market confronts traders with all of the challenges associated with an OTC market, i.e. a bilateral, non-transparent market. Prices are hence frequently only quoted indicatively (i.e. as non-marketable prices), and sometimes they're missing altogether. This can make it very difficult to thoroughly calculate an entire trading strategy, as it always depends on various market makers and their price quotes. The problem once again illustrates that market experience is a paramount requirement for the accurate interpretation of price indicators and market sentiment.

The generally published gold price principally refers to the location London and a minimum fineness of .9950. If a price for Loco Zurich is to be calculated, for instance, a market maker needs to include the theoretical consignment surcharge from London to Zurich. But sometimes important gold hubs are also traded without any surcharge. But how can it be that some trading locations are traded more expensively, while others have no surcharge? The answer is obvious: The fluctuations result from supply and demand. Large market makers usually hold gold stocks in all major trading centers. It is therefore a frequent occurrence in bilateral price trading that gold is bought up for every location and is then on average principally available at all storage locations. But as bottlenecks sometimes do occur, the prices may be subjected to different market rules.

There is still a basic difference between electronic arbitrage based on unallocated gold accounts and the actual, physical allocated gold arbitrage. The latter is to be viewed as a purely theoretical operation, as a physical consignment is usually so cost-intensive that possible profits are used up.

Information Efficiency in Cash Arbitrage

The first basic question is: "What is the price determination based on?" An examination of market operations quickly reveals that the price published by Reuters, for instance, forms the basis for virtually all gold traders who are active around the globe. And a closer look at this price source shows that the quote is comprised of indicators from various banks. The price quotes are not binding and are sometimes published by the banks with significant delays and spreads whose extent exceeds the usual market norms. This demonstrates that there is a difference between actual prices and those quoted to the general public. The quotes are also conditioned by the trading periods of active and less active markets and the situation of the market makers. The fairest and most liquid prices in the spot market are certainly quoted in the early hours of the London gold trade. All the market makers represented in the LBMA are active at this time of the day and the continuous communication and interaction results in a fairly efficient gold price determination. But this is permanently subject to change during the rest of the trading day.

Another barrier to information and price efficiency is the price determination per se. There is no relevant electronic exchange for spot gold. This means that prices are determined via telephone or Reuters quotes (i.e. bilaterally). That this procedure is, in the age of the Internet and electronic trading platforms, at a disadvantage compared to "newer" markets is obvious. Another factor contributing to price efficiency, apart from reference price and price determination, is obviously price supply. Not all market makers act or quote alike—rather the opposite is true. Each market participant contributes something uniquely individual to the market.

The function of a Loco arbitrage is clear. But this leaves the question: Who exploits it or, respectively, prevents the occurrence of market imbalances? It is often the smaller trading establishments and proprietary desks that perform various price comparisons, because these trades are usually not marketable in large volumes and result, in relative terms, in insignificant proceeds. Banks and gold market clients usually acquire quotations from several gold traders before they close a deal, which often ensures that the market price is relatively well aligned.

Arbitrage traders are very useful, particularly in the gold market, because they always turn up when prices are "wrong," be it because the wrong prices were quoted or because a market maker has a different price opinion and therefore determines a different quote. Seen from this perspective, arbitrage brokers prevent excessive price spreads. In addition, some banking houses are by now globally positioned in the gold market and hence in a position to standardize information transfers, and therefore price determination, within their own organizational structures.

INTEREST ARBITRAGE AND SPREAD TRADES - ARBITRAGE IN FUTURES MARKETS

The function and purpose of futures markets have already been explained in the first chapters. The information presented there is required in order to be able to examine the function of interest arbitrage and interest curve trading strategies more closely.

FORWARD FUTURE ARBITRAGE

Forward markets and futures markets differ in their individuality or standardization, respectively. But their basic concept remains the same. Thus, the existence of the various entrants in both markets opens up the first arbitrage opportunity between the OTC forward market and the standardized futures exchange trades. This arbitrage opportunity is somewhat analogous to spot arbitrage, but also features a few peculiarities of its own, as the following example shows:

The March 2005 contract maturing on February 28 and to be delivered on March 10, for instance, is currently traded at $452.40 USD. The current spot reference price is $448.00 USD. The swap or contango, respectively, hence lies at $4.40 USD. That's 2.15 percent per annum, in relative figures. (92 days between maturity and spot valuation). Now all you need for comparison is a forward price quote from a market maker. The OTC forward market's quotes are in percent per annum, in this case 2.00 to 2.10 per annum.

The following strategy ensues: Acquire the OTC forward at 2.10 percent and sell the future at 2.15 percent. The initial profit is five basis points, 0.05 percent resulting from the interest rate spread.

The positions entered into are a futures contract in the London market and a short position in New York.

Irrespective of where the prices go next, the profit of five basis points is guaranteed. The one thing to be considered is of course the geographical distance between the contract obligations. The short contract needs to be delivered to a certified COMEX depot, whereas the gold from the OTC futures acquisition needs to be received in London. As we can see, the trade results in a physical and logistical challenge. The gold bought in London needs to be transferred to New York via a swap. The profit is hence consumed by the swap.

INTER-MARKET SPREAD

An inter-market spread consists of a long position and a short position in contracts of identical maturity at different exchanges. But because the two largest futures exchanges, the TOCOM and CO-MEX, do not overlap chronologically, this type of arbitrage/trading strategy is only possible at smaller exchanges. A classic inter-market spread can therefore be achieved between TOCOM in Tokyo and CGSE in Hong Kong, as the following example illustrates:

A future contract for gold with delivery in June 2005 is bought at CGSE in Hong Kong at a futures price of $453 USD, while simultaneously a contract with the identical maturity is sold at TOCOM for $454 USD. Both contracts should theoretically have the same price—and that fact exactly is the motivation for the trade: the approximation of both contracts to their fair value. Once both prices align again, the positions are closed again. This trading strategy is obviously not a classic arbitrage as it comprises a certain risk regarding expectations of the spread being closed.

During the period of the position being open, there is a currency risk resulting from the TOCOM contract quoted in Japanese Yen and the CGSE contract, which is quoted in HK-Dollars. This risk should theoretically be safeguarded against by currency hedging instruments, but a detailed description of this procedure would lead too far at this point.

The spread's profit results from the price spread between the exchanges, multiplied by the total number of contracts entered into and the underlying number of ounces per contract, minus transaction costs for the execution of the four trade positions and possibly the added expenditure of a currency hedge. This creates conversion problems between various bar sizes, currencies, finenesses and units of weight. The following table serves to clarify these:

Date	CGSE Price	In US-Dollar Ounce	TOCOM Price	In US-Dollar Ounce	P&L Position
24.2.2005	Contract in Tael (10) 4678,59 HK-Dollar	Long (Kauf) at 453 US-Dollar per ounce	Contract in Gramm und Yen (1000 g) 1234,4 Yen	Short (Ver-kauf) at 454 US-Dallar per ounce	1 US-Dollar per ounce
25.2.2005	4689,00 HK-Dollar	Short (Ver-kauf) at 455 US-Dol-per ounce	1233,80 Yen	Long (Kauf) at 453 US-Dollar per ounce	0 US-Dollar per ounce
Profit/ Loss		2 US-Dollar per ounce		–1 US-Dollar per ounce	–1 US-Dollar per ounce

Inter-market Spread
Source: *Reuters, Bloomberg, Author's own calculations based on CGSE, TOCOM*

As the contracts at both exchanges are based on different metal weights, a balance between the differing weight units can only be achieved by entering into positions with divergent volumes in the respective contracts. At TOCOM, each contract buys or sells 1,000-gram gold, whereas a contract at the CGSE represents 10 Tael. The overall profit in this case is $1 USD per ounce before cost—it obviously needs to be leveraged by a suitable volume to make the yield worthwhile after deduction of costs.

Spread traders try to close their open positions as quickly as possible in the face of the spread realigning with normal market conditions. In this case they also try to avoid delivery and settle both contracts at the exchanges before they reach maturity. Because if the contracts are delivered, a rather complex continuation of the trade ensues as described above.

The risk of the spread not closing can be reduced insofar as both positions are simply held until the contracts' final maturity, hence triggering the underlying physical delivery. But a physical delivery should really only be considered if profit costs exceed delivery costs.

TIME SPREAD

This trading strategy is also sometimes called intercommodity spread trading and involves entering into a long position and a short position for the same precious metal at the same exchange but with two different maturity terms, for instance, by purchasing an August 2005 contract and the sale of a December 2005 contract. An arbitrage profit comes about if the price spread between contracts exceeds or undercuts the interest costs, the so-called cost of carry. The cost of carry theoretically consists of transaction, storage, transport, insurance and financing costs. But in real life, only lending yields/costs and financing costs are usually set off against each other as cost of carry. The calculation formula for these expenditures shall be explained in the following pages.

The gold futures market has been a contango market for several years now, meaning that forward prices exceed spot prices. The difference is a markup, the so-called premium. (If forward prices are lower than spot prices one would speak of backwardation and a discount, respectively.) The contango can be compared to an interest structure curve, similar to bonds. The larger the distance between both contracts, the higher the cost of carry will usually be. If this curve ascends more than normal between the contracts, there might be a chance for arbitrage. It is hence of paramount importance for the trader to permanently know what the fair value of an interest curve should be. He can always calculate the fair value by a simple formula, but he still needs to keep a watchful eye on market conditions, which are subjective. He can then, for instance, compare the theoretical interest curve with the one actually being traded in the market and, should a significant divergence occur, check the causes. If he then decides that the price difference between the fair value and the price traded in the market is not justified, he can base a trade on this market situation.

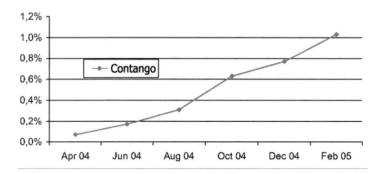

Contango at COMEX for Gold - Premium in Percent of Spot Price
Source: *Datastream, Author's own representation*

The illustration shows that the price rise between August and October exceeds that of other months. This phenomenon is not explicable by the theoretical price. If the trader wishes to exploit this situation, he needs to sell the contract deemed expensive and buy the "normal price" one. As I have explained in the context of spread trades that the trader expects that the market will recognize and retract this price shift at a given point in time. Should the price difference between both contracts still be higher than the actual cost of carry, by the end of term, the trader simply has the August long contract delivered. He deposits it until the October contract matures and delivers the position resulting from the August contract for the October contract. The profit in this trade would be the price difference between the contracts minus the cost of carry actually incurred.

Intercommodity Spread: Premium vs. Cost of Carry
Source: *Author's own representation*

But if the premium is lower than the cost of carry, the spread trader sells the August contract and buys the October contract against it. If the spread does not balance until maturity of the August contract,

GOLD

the trader needs to lend the gold in the period between August and October to cover his obligations according to the August contract. The October long contract delivers the gold, which needs to be borrowed up to this point in time. The cost of carry is hence calculated "backwards." The gold is borrowed for one month and that is what the lending interest is paid for. The proceeds from the sale of the "August gold" are invested at USD short-term interest rates.

The spread between cost and yield compared to the premium traded creates the profit.

Intercommodity Spread: Premium vs. Cost of Carry
Source: *Author's own representation*

The only risk in time-spread trading is that of the cost of carry changing, i.e., the risk of interest rate adjustments. The term cost of carry is similar to bond trading and usually includes financing costs. With gold there are some additional factors. But the cost of carry is still significantly lower in the gold market than in other commodities markets because the storage and administrative costs created by raw materials, like cattle or grain, tend to be somewhat higher. The reason for this is the homogeneity of the raw materials, among other things. Gold can be stored as a homogenous good, whereas soft commodities, i.e., agrarian produce, not only need to be stored in special cold stores or warehouses but are also not homogenous. Another difference is that large portion of gold sector activities is nowadays carried out electronically, thereby reducing the need for physical transactions, if not dispensing with them altogether.

The current gold market has changed insofar as gold lending rates have hit a historic low and tend to partly be negative at the short end of maturities. Lower lending rates mean a higher contango. The contango thus increasingly approximates financing costs, usually consisting of the USD interest rate for the refinancing. The mathematical formula is:

$$F(T) = S[1+r(T/360)]/[1+r^*(T/360)]$$

The theoretical future/forward price results from a function of the spot price (S), the lease rate (gold lending rate) (r*) and the USD interest rate (r) over time (T), representing the number of days. This is the formula for calculating the fair value; its result can be compared to the contango actually being traded. Both "price bands" are usually very closely aligned because the market would be highly inefficient otherwise.

A comparison between the mathematical contango with the traded contango in the following illustration reveals that a significant arbitrage would have been possible several times over. But one generally needs to consider that, even in these periods, the trades' realizability needs to be checked first, as does a relevant volume.

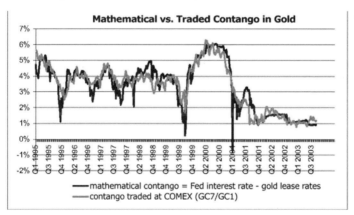

Mathematical vs. Traded Contango
Source: *Datastream, LBMA, Bloomberg, Author's own research*

The traded contango in this illustration equals the premium between the one month future and the one year future. The mathematical contango was calculated using the FED interest rate and the lending rate for gold, valid at the relevant times. For the lease rate, a large number of different market makers were queried to gain a better overview.

GOLD

As we can see, the market traded a higher premium than necessary in purely mathematical terms in 1999, for instance, when the
I. Washington Agreement on Gold came into force. The alignment
occurred within a few weeks. The trading strategy demonstrated
above would have made a profit then. As current lease rates for
gold are very low, the mathematical and traded contango is quite in
line with financing costs, and divergences can only be spotted rarely
or are only rarely efficient enough to be exploited for a profit.

Another consideration is that the calculations performed for our
example refer to the spread between the one-year contract and the
forward market, which sometimes suffers from insufficient liquidity. The gold market at COMEX is usually only liquid during the
front months (i.e. the months nearest to the contract), whereas the
back months (i.e. contracts with a longer term), are usually only
traded sporadically. This is why the OTC forward market should
also be included in this type of deliberation. Because it is the OTC
forward market that puts one in a position to trade larger volumes.

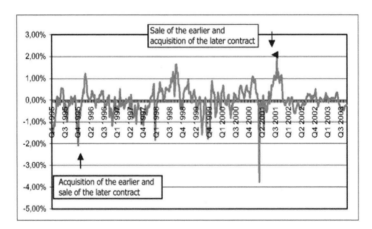

Spread Between Mathematical and Traded Contango
Source: *Datastream, LBMA, Bloomberg, Author's own research*

If you'd like to monitor this spread strategy over a longer term,
it would make sense to record the spreads of both interest-rate
curves. This way, you can always immediately see if the market is
traded above or below the contango.

If the spread is zero, the premium exceeds the cost of carry; therefore, the earlier future is bought and the later one is sold. If the difference line rises above zero, the earlier contract is sold and the later one bought. If the line goes back to zero, the positions can be closed again. The following rule applies: The higher the divergences, the higher the chance of a profit.

To further illustrate the procedure of interest-rate curve arbitrage, let me show you two example calculations that occurred in the mid-1990s. In the first example, the mathematical contango is lower that the traded contango, i.e., the traded premium exceeds the theoretical cost of carry.

In the first example, the 1995 December contract is bought and the later contract for December in the following year is sold.

	Mathematically Traded Contango	Contract December 1995	Contract December 1996	Profit / Loss
1.12.1995	– 2,07 Percent	Buy at 386,10 US-Dollar	Sell at 398,4 US-Dollar	+ 12,3 US-Dollar
8.12.1995	– 0,49 Percent	Sell at 389,9 US-Dollar	Buy at 401,9 US-Dollar	– 120,0 US-Dollar
Profit		+ 3,8	– 3,5	+ 0,3

Time Spread - Short Holding Period
Source: *Datastream, Author's own calculation*

The contango has been already significantly reduced within a week of entering the position. The profit before transaction cost is $0.30 USD. But as the transaction cost for all four transactions total approximately $0.60 USD per ounce, this arbitrage would not be profitable in practice. The transaction costs refer to the expenditure required for a COMEX broker (in this case FIC). To facilitate a profitable trade, the difference needs to retract even further.

But what happens, if one wishes to hold a position for longer? The earlier contract matures after about one month and needs to be rolled. This means that the contract due needs to be brought into the next contract shortly before reaching maturity. This can either be done by selling the contract and buying the new one or by buying the roll per se. The roll represents the costs of the situation described above, but simplifies the trading. If a roll is bought in this case, the position is prolonged for a further month.

In the second example below, the trader just holds the position a little longer, meaning that the first contract is rolled in the expectation that the spread will then approximate. The costs of the roll are not considered in this example.

	Mathematically Traded Contango	Contract December 1995	Contract December 1996	Profit / Loss
1.12.1995	– 2,07 Percent	Buy at 386,10 US-Dollar	Sell at 398,4 US-Dollar	+ 12,3 US-Dollar
12.1.1996	– 0,06 Percent	Sell at Jan-96-Kontraktes 396,7 US-Dollar	Buy at Jan-96-Kontraktes 406,6 US-Dollar	– 9,9 US-Dollar
Profit		+ 10,6	– 8,2 Percent	+ 2,4 US-Dollar

Time Spread - Longer Holding Period
Source: *Datastream, Author's own calculation*

As the mathematical contango in January was even higher than the traded contango, closing the position earns a significantly higher profit. In this case, a trading profit of $2.60 USD per ounce is set off against costs of only $0.60 USD per ounce—it is therefore easy to understand that the profit after cost of $1.80 USD makes perfect sense of the trade, relatively speaking. If we assume that the trade is executed with the usual trade volumes, for instance 30 contracts per maturity term, the profit would be $5,400 USD (30 contracts, times 100 ounces, times $1.80 USD profit).

It is never required in futures trading to pay the entire sum represented by the contract, but only a part of it, the margin. Different margins apply to different contracts and exchanges and also vary depending on whether the transaction is a speculative purchase/sale or a commercial transaction, i.e., a hedging strategy. The exchanges principally charge small or even sometimes no margin payments at all, which means that large volumes may be leveraged with relatively small capital.

CASH AND CARRY ARBITRAGE

This classic arbitrage fuses the elements spot trade, lending market and forward market into one single strategy. The principle might be simple, but the execution is somewhat complicated.

As I have shown in previous chapters, the forward price is composed of a combination of USD interest rates and gold lease interest rates. Traders are hence always able to see whether a forward price is, relatively speaking, too high or too low, by simply perusing the spot price.

To simplify this process, the markets have developed EFPs (Exchanges for Physicals), which record the contango, i.e., the spread between spot and forward price. If we, in purely theoretical terms, assume that all market entrants act the same way, there could never be any price differences. But reality is quite different, primarily because the cash market is mainly based in London and the most important futures market is in the United States. All information and market trends are hence evaluated quite differently, meaning that prices can display different trends in the short term. But although this is the case, the afternoon trade in London has in recent years been so strongly focused on New York's COMEX that most market makers first consult COMEX as a price basis before quoting prices for the London market.

It is nonetheless interesting to examine cash and carry arbitrage more closely. The initial motivation for this trade is if the trader deems the futures price to be too expensive in comparison with the spot price. This means, according to the trader's price components, another price should result. If this is the case, he sells the future contract in New York and simultaneously buys spot gold in London. Now the trader needs to lend out the gold he bought in London until the New York contract matures. The money he needs to pay for the gold will be reimbursed in the shape of the contango of the gold contract sold. The price difference may hence be rooted in three different components. The trader is finally required to transfer the gold to New York via location swap. This creates expenditure that needs to be included in the trade calculation.

This type of arbitrage is only rarely possible these days. It can also be quite work-intensive to execute this type of trade accurately. In addition to this, the interest currently paid for gold at the short end is negative—the trade is therefore not profitable in this day and age. Yet in the 1990s, this was a relatively frequent transaction. Times do change.

CARRY TRADE

The carry trade has been regularly executed by several market entrants in recent years. But it ceased being interesting after the gold market climbed out of its slump in 2001. The trading strategy as it is described here is not an arbitrage trade in the classic sense because it includes an open risk position. But it is still important to understand this strategy, because it not only shows why the gold market was a bare market for many years, but also demonstrates how market participants still tried to exploit it for their own ends.

The carry trade starts with a gold lease from a lender, mostly central banks or bullion banks. The leased metal is sold in the spot market at the current price. The Dollar yield (or the yield in other currencies) is then invested in bonds or financial instruments earning a similar interest rate, the expectation being that gold prices will drop and that the metal can be bought back in the market later at a more favorable price. After the gold has been bought back in this manner, it can be returned to the lender. The lease rate is usually lower than the financing cost, which means that there is usually a profit to be made from the interest spread.

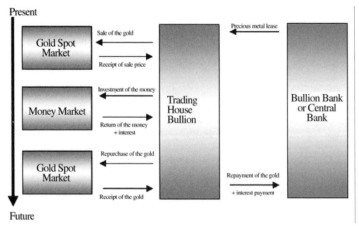

Carry Trade
Source: *Author's own representation*

This transaction's profit mainly depends on future gold price developments as it is determined by an open short position. In the reverse case, i.e., if prices go up, the trader stands to suffer a significant loss. The profit is calculated in the following manner:

Sale proceeds at spot price

- + interest from USD-investment
- - leasing interest for the gold
- - purchase amount at future spot price (the only as yet unknown component)
- = profit/loss

As I mentioned in the beginning, this trade was very popular with some trading establishments and speculators in the 1990s. They exploited the weakness of the gold market and went short, covered by borrowed gold, which resulted in additional pressure in the gold market. But the turnaround in the gold price curve ushered in the end of this trading strategy.

REVERSE CARRY TRADE

The reverse carry trade has virtually the same structure as the carry trade, except for one significant difference: Instead of the gold, a financial amount is borrowed to buy gold and lend it to banks or industry. The trader's assumption being that the interest rates for gold are higher than the interest to be paid for USD.

This trade also results in an open position for the trader, this time concerning price drops, because he has to buy the gold first in order to be able to lend it out. And he later has to sell the gold again in the market in order to repay his USD-credit.

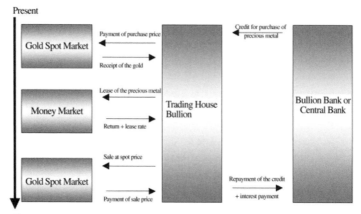

Reverse Carry Trade
Source: *Author's own representation*

The profit or loss, respectively, is calculated in an analogous manner to the carry trade.

Neither of these trading strategies—carry trade and reverse carry trade—is an arbitrage strategy. They are still executed by international trading houses regularly if and when the individual trade looks promising. With this type of trading, it is of primary importance to closely follow lease rates and market trends.

CONCLUSION: TRADING STRATEGIES IN THE GOLD MARKET

A real insight into the gold market can usually only be attained from insiders, traders or analysts. The relationships between mines, banks and industry often only become clear once an observer understands the role played by each individual market entrant—role, in this case, meaning in terms of their respective activities at the exchanges and in the OTC markets.

Mines, for instance, basically act as sellers in the gold trade. But new financial products and strategies permit the mines to simultaneously act as buyers of forwards and as sellers of options. One of the classic mine trades is for instance the purchase of so-called bandwidth options, also known as zero cost collars. In this trade, the mine buys an option, entitling it to sell its gold at a guaranteed price level, in exchange for the obligation to nonetheless sell its gold up to a certain level. In this strategy, the mine buys a put and sells a call to finance the put.

The fixing is an instrument intensively used by mines. It is usually of little interest to hedge funds which preferably use COMEX to develop or shrink their positions, reason being the market's liquidity and transparency. Here, a hedge fund manager can monitor the market price at which he ordered, because prices are distributed via various data services. But he can also roughly assess the extent of the market's liquidity, and if and when he will be able to make arrangements for his positions in this context. Hedge fund managers especially often try to exploit "thin" market phases in order to trigger market movements with their transactions. This is, of course, quite contrary to the purposes of the fixing: to place the largest orders possible (on both sides) without attracting a lot of attention in the market.

An inspection of the global gold market reveals a variety of hierarchies and focuses: The central point is occupied by Asia with its two gold centers, Hong Kong and Tokyo, with relatively short trading periods—a market that sticks to itself, so to speak, far from the European market which only wakes up incrementally some time later. Europe itself is divided. It dominates the physical market in the morning hours while simultaneously acting as a global price fixer, but as soon as COMEX opens its doors in the early afternoon (European time), the significance of the European market recedes once more.

The afternoon fixing is hence the most important price factor globally. At the time of the afternoon fixing, the largest gold trading nations are active in the market, as are various interest groups: mines from South Africa and the United States, European banks and hedge funds, as well as the major central banks. This is also the reason why the afternoon fixing features the largest volume. Some market participants in this respect even speak of the only true gold price, because it is traded and accepted by the mines and by the other market entrants. It is also often used as a primary price reference, whereas the morning fixing is mostly used by the Australian mines and European banks. The Australians usually operate via their London branches.

GOLD

Once the COMEX trade closes, the gold trading day is all but over. This still leaves the physical market in New York with its spot trade, and the global trading platform, NYMEX-ACCESS, is also still active, but the volumes wither to insignificance.

This raises an important question: All traders talk of profits—but what about the cost? The trade creates numerous expenditures, for example in the shape of broker commissions and exchange fees. This applies to the gold exchanges in Japan, Hong Kong and Istanbul as well as to New York. But the OTC trade is usually bilateral, with only a few brokers active in bullion broking. In the London market, the major bullion brokers are GFI and Garban Intercapital. In the Swiss market - still from the time of the Zurich gold pool—it's the Premex, now trading under the name Cosmorex AG. The cost of OTC bullion broking depends on the broker but usually ranges around $100 USD for 2,000 ounces of gold. Besides spots, the brokers also quote forwards and options.

In conclusion, a brief summing up of the international gold trade: The market is unique. It is a small market, the traders all know each other—their numbers not being that overwhelming. The conferences and meetings are very traditional just as the market itself is extremely aware of its heritage. It has a special aura, bordering on the mythical. It is the physical component, as represented by the raw material, that lends the gold market its special allure. Not everybody is in the know. Only very few people understand the system, which is closed in itself.

But the trade would probably still be quite boring, despite this exclusivity, if gold were to be globally traded in ounces and USD on one platform only. But the variety of the gold itself and the market impurities resulting from this lend the market its unique excitement. The very inefficiency of information can be one of the main motivations for gold trades and arbitrage. Only this inefficient information facilitates a global market perspective and trade. Gold might well be a homogenous good in itself, but because of the multitude of variations, it is everything else in terms of trading. The result is niche markets and market peculiarities.

A comparison between the current gold trade and the situation a few decades ago shows very clearly that the market has definitely developed in a positive direction. While the gold trade was illegal in countries like India and South Africa at the time the process of liberalization has made great strides in the meantime. Only in China and Vietnam do certain restrictions still apply to foreign traders. The liberalization is not all-encompassing here—and it is in exactly such cases that market inefficiencies are created, because markets that are not subject to liberal and free economic principles often quote different prices. And this is eminently visible in gold. The gold is often traded at a significant premium in restrictive countries, resulting in flourishing smuggling operations and illegal markets.

If the gold market should ever be liberalized globally it would be the end of gold smuggling. Then, the days of premiums and discounts will be over and gold will only have one uniform price: the fair price!

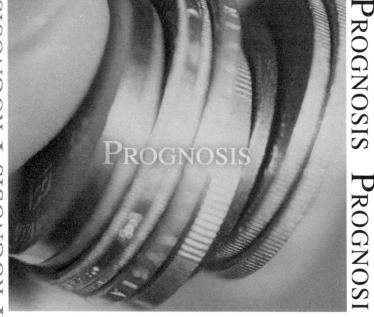

-IV-

Any market, be it for gold, shares or bonds, is characterized by its market entrants. Their customs, standards and trading habits have either developed over the years or were enforced by statutes.

This also applies to the gold market, one of the most traditional markets of all. Its current structures have been in existence for decades. The market entrants have, for the most part, the same, with the exception of the trading establishments that might have a different name nowadays. Gold fixing, London Good Delivery bars, LBMA—these are all terms that show that the gold marked grew organically.

This is not to say that the future ahead is entirely golden. In this age of globalization, the market has become larger and more efficient, but the pressure has also increased, and massively so. The managers of today's investment banks decide on the gold markets future and development purely on the basis of profit and loss calculations. Heritage, history or reputation does not really enter into sober figure crunching. The latest example of this is the private banking establishment NM Rothchild which, despite a heritage reaching back centuries, had to retire from the precious-metals trade due to low profitability. From a gold trader's perspective, all one can say is that it can be quite sad when tradition and cultural heritage are simply thrown overboard in such a manner.

Another thing a closer look at the gold market reveals is the fact that the traders are, on average, older than in other product classes. New faces have only been entering the market again in the last couple of years. In the years before that, more personnel left than came, because the NASDAQ, the Internet boom, etc., were more attractive for young traders. This is a phenomenon that also affects exotic raw materials, like wheat and palladium, and to a much larger extent.

Let us take a look at what has changed in the market and who the groups of market entrants are today. These groups have changed drastically, particularly by growing in size, especially if you compare the situation today with that of 20 years ago.

There are basically four groups of market participants:

1. Hedge funds, CTAs, speculators
2. Central banks - public sector
3. Industry (demand and supply)
4. Investors (long term oriented, strategic)

In their basic structures, these groups have been around for decades, albeit in smaller versions. But a closer look at the central bank sector, for instance, reveals that this group of market entrants currently almost exclusively acts as a tender (i.e. a supplier of gold) in the market. The net volume of gold sales by central banks in 2003 was 600 tons, which is the approximate equivalent of a quarter of the annual global gold production.

The following sections will examine the individual groups of market entrants more closely.

HEDGE FUNDS - CTAS - SPECULATORS

What do all these market participants have in common, you might wonder? It is the absolute return. A hedge fund tries to generate absolute returns without putting these in relation to, let's say, a benchmark. The instrument used to generate these absolute returns is strictly secondary—return is return.

Speculators are generally not much different in this respect; what's important is the yield—not the performance compared to some benchmark.

But what about the CTAs? CTAs are more of an American phenomenon than a European one. CTA stands for Commodity Trading Adviser, and the name basically explains what these companies do for a living. They supply consultancy services to clients and manage client assets, primarily in raw materials markets. Their trading decisions and advice are often based on computer programs or technical models. As it is customary in the USA to enter into raw materials investments via a CTA, their asset management mandates are rather sizeable which lends the CTAs significant sway in the markets.

My broker constantly tells me, "Chris, never ever underestimate the power of the CTAs." The official COMEX figures of the CFTC, the Commodity Futures Trading Commission, clearly illustrate his point. The CFTC categorizes all market entrants into commercials and non-commercials, i.e., into traders who have a fundamental interest in the market (such as gold mines or jewelry manufacturers) and in speculators who "only" trade on an opinion.

CTAs often trade based on technical marks and signals, which also explains why the classic chart method can be used more successfully in the raw materials market than in any other markets. Persons wishing to place money with a CTA have two basic options:

They can grant a purely advisory mandate for which trading advice will be given, or they can commission a management mandate comprising the management of assets according to various criteria.

The group of speculators includes individuals speculating in the market and the proprietary desks of banks. These might be insignificant individually, but as a whole the market segment moves a lot of volume. The methods and manners employed by this group in the market have already been partly described in chapter 5. It would be wrong to assume that the mass of speculators trades via complex arbitrage trades. In fact, they mostly place classic long/ short/outright bets.

But it is still a great challenge, and some kind of game, for speculators to perform a squeeze in raw materials markets. A squeeze is exactly that! It aims to, for instance, create an artificial shortage of a raw material in order to heighten demand. The normal reaction in such a case would be that buyers who urgently require the good, for instance because it is a significant component of a production chain, are prepared to pay almost any price for it.

A squeeze is basically very hard to bring about in the gold market because the market depth and the volume of the already existent (already mined) gold are quite large. But it is imaginable in theory, as the following example from the silver market shows, which, although it functions in a manner similar to the gold market, comprises a much smaller volume.

In the early 1980s, a group of speculators surrounding the billionaire Hunt brothers tried to squeeze, or corner, as it is also sometimes called, the silver market. To this end they bought up silver contracts of any maturities in forward markets. All investors who had gone short on these contracts had to deliver the silver to the brothers at the end of their terms, insofar as the brothers wished a delivery. And that exactly was their plan: to have each contract delivered physically. The aim was to create a silver shortage in the market. They achieved this by purchasing massive amounts of silver in the spot market and literally absorbing any material that was available via leases. Any investors who were short in the futures market and didn't actually own the silver (which is usually the case with speculators, in contrast to the mining or recycling industry) were hence forced to physically buy it. But more or less all the silver in the market had already been bought by the Hunt brothers. The result was that the silver price very rapidly increased from $6 USD to over $40 USD. The Hunts had done their homework, but their concept was a flop in the end because the futures exchange of the time quickly decided to change the market conditions in such a way that a squeeze would no longer be possible. The fact that the exchange itself briefly intervened in the market in order to serve its own interests by changing the rules seems just as worthy of condemnation as the intention of a silver squeeze in itself. The silver price thereafter retracted to its original level and the Hunt brothers had to accept a crunching and defeat that was not without consequences.

The parallels to the gold market are obvious. But there is one important difference between the gold market and the silver market at the time of the Hunts' speculation. There are large stocks.
Central banks, some asset management firms and hedge funds own sometimes truly massive gold positions that could be sold incrementally if the prices rose, which would be the case in a squeeze. But it is still important to understand how raw materials markets function if one wishes to exploit the character of such a market for personal gain.

Speculators such as the CTAs and hedge funds are by no means always the bad guys trying to benefit from other people's misfortune. As a matter of fact, these markets wouldn't be able to function without them. This group supplies the market with liquidity and transfers risks. If a market would, for instance, only admit professional groups, the following would happen: A mine produces gold and wishes to sell it as soon as possible while the jewelry manufacturer actually requires gold, but only in three months' time. This would lead to a so-called time mismatch. And even if they pursue the deal further they might discover that the gold volumes don't even match: The gold mine would like to sell several tons of gold, whereas the jewelry wholesaler only requires several hundred kilograms. This additionally creates a size mismatch.

This is where the speculators come in: They act as intermediaries who are prepared to run a risk if they see a profit in it from their perspective. They thus create more liquidity and a deeper market penetration, although this may be subject to higher fluctuations in particular market phases. The advantages and disadvantages are closely interlinked, but the market wouldn't function without these groups. The most recent gold market trends show that it is currently attracting further representatives of this group and is hence able to grow. Retail products, such as certificates and warrants and the growing number of CTAs since 2001, are positive signals for the market and if this development continues, the framework for a positive gold price cycle will remain intact.

CENTRAL BANKS - PUBLIC SECTOR

Central banks and public financial institutions like, for instance, the Kuwait Financial Authority or the Singapore Financial Authority have a key role in the gold market. In no other market is this group this reserved, almost mysterious, and simultaneously as massively influential in price terms. Why should this be the case? To answer this question, one needs to look back in history over several decades.

Until the failure of the Bretton Woods agreement, the major currencies were backed in gold, at least partly. This gold backing meant that central banks planning to circulate new money were forced to finance it in gold. The result was a massive increase in the central banks' gold stocks. The USA, for example, today hoards a gold treasure of over 8,200 tons. The Federal Republic of Germany comes in second place with roughly 3,400 tons, closely followed by France, Italy and Switzerland. The more sizeable the official gold stocks, the more trustworthy a currency used to be.

But in recent years more and more voices could be heard maintaining that it was hardly intelligent of the central banks to hold gold as a strategic currency reserve. That the gold didn't earn any interest was one of the arguments used by these individuals (which is only partly true, as gold can be lent out and hence earn perfectly good interest), and the other was that it served no purpose and was indeed a "relic of barbarism," to put it in Keynes's phrase. Nothing more than a yellow stone, so to speak.

The result was that some central banks, for instance of France and Great Britain, have massively reduced their gold positions in the last ten years. This strongly affected the gold price until it finally, in 1999, reached a low of $250 USD per ounce. In reaction to this, some of the major central banks got together to hammer out an agreement stipulating that gold sales should henceforth be executed in a more coordinated and market sensitive manner: the I. Washington Agreement on Gold, which was also signed by Germany. The agreement decreed that participating central banks are obliged to sell no more than a total gold volume of 500 tons per annum in the market. The result was a massive price rally because the uncertainty about how much gold the central banks would sell had been abolished. The original agreement expired in September 2004, but was extended for another five years. The conditions remained the same, except that the maximum volume was raised from 500 tons to 2,500 tons.

But there is still the odd central bank here or there which sells its gold or even buys fresh gold, as was demonstrated by the Argentine Central Bank in August 2004 when it purchased over 50 tons of gold. But no large scale purchases have been recorded to this day.

Let me at this point take a brief look at my domestic market by considering the activities of the Deutsche Bundesbank. The German central bank had initially reserved the right to dispose of 500 tons of gold in the II. Washington Agreement on Gold. But on December 20, 2004, it announced that it was not planning to peruse its sale option in the first year (with the exception of eight tons for minting). This does not mean that the Federal Republic intends to hold its gold indefinitely without ever selling any, but rather that no clear consensus on the sale has as of yet been reached between the government and the management of its central bank.

A discussion of gold in the context of central banks always demonstrates gold's role as a currency with particular clarity. Central banks, most of all, tend to perceive gold in a classic monetary role. The metal is, in this context, no longer viewed as a raw material that is being mined and turned into jewelry. Just as some parts of the population regard gold as a safe asset and emergency reserve—which is most of all the case in India, as we have seen—some financial authorities and central banks continue to hold gold as a reserve, their arguments being emergency reserves, trustworthiness and diversification—all in all pretty convincing arguments.

INDUSTRY

The industry is basically split into producers and buyers of gold. The production side is made up of the mining industry, i.e. all the mining companies scattered around the world, as well as the refineries and separating works, some of which process old gold or recycle gold from discarded electronics equipment. The buyers in the gold market comprise jewelry manufacturers, wholesalers and again some refineries and separating works.

A closer look at a gold mine's finance department, also known as the gold mine's treasury where most gold sales are managed, gives a vivid impression of how the mining industry relates to the gold market. The treasury's most important activity is finding gold buyers in the market. To this end, it is attached to the market via banks that have good knowledge about the available instruments and the markets themselves. The market entry of the company depends on the strategy to be pursued. In the last three years, since the gold price has been rising again, the companies have become very reserved concerning hedging transactions. This means that they sell their gold in the spot market, mainly via the fixing, because this is where the greatest liquidity is created. Of the two fixings described in this book, they usually prefer the afternoon fixing because it occurs during the opening times of COMEX, and the market is at its most liquid then. (This is also the reason why the afternoon fixing price is weaker in a statistical evaluation than the price of the morning fixing. Certain unlikely arguments often proposed by the international gold price conspiracy community that the gold price is being manipulated in the USA are not explicable on these grounds!)

This internationally valid price reference is used to calculate the mine's profitability. Sometimes the mines also utilize opportunities to balance individual, smaller spot positions in the market with excess production without employing the fixing. But if we look back 10 years or so, it appears that mining companies acted quite differently in the market. Hedging was the fashion then because there were opportunities to exploit higher prices in a falling market. This is why some of the companies also employed the possibility of future sales and frequently depended on instruments such as forwards, leasing and options.

This fact should not be misinterpreted: There is still some hedging going on today, but on a much reduced scale. Small portions of the production are still hedged because hedging transactions are still required due to bank demands concerning project financing. In a market where the gold price is falling, hedging makes eminent sense and is supported as a strategy both by shareholders as well as entrepreneurs. But rising gold prices can have drastic consequences for a company's profitability, which is what happened in 2004, when a well-known Australian mining company, Sons of Gwalia, had to apply for protection against creditors because the hedging positions already entered into had taken a dramatic nosedive. A de-hedging process has hence been observable in the market during the last three years. De-hedging means that companies holding hedging positions simply wind them up again. Due to this termination, the mines need to enter the market on the buying side. The rise in gold prices during recent years is thus also attributable to the mining industry's increasing demand.

The demand side is another world entirely. The jewelry industry, with its main buyer India (over one-fifth of annual production goes to the Indian jewelry market), is less active in hedging instruments than the mining industry. It usually acts purely as a buyer in the market and acquires gold depending on its requirements. India is a classic example of this. As I have described in chapter 4.10, the Indian subcontinent usually buys in cycles caused by the Diwali Festival of Lights and the wedding season. During these phases, an enormous amount of physical material is shifted, albeit at the current market price.

Refineries also act as buyers in the market, but they usually only execute client orders and limit their risk positions by hedging transactions.

So what can we say in conclusion about the future of this group of market entrants? It is basically a passive group, clearly composed of price takers. Positions are usually managed and executed without incurring significant risk positions. But this can and will change. Even if hedging does not represent a mine's current company policy, price situations or company decisions may well arise that put it firmly back on the agenda. For this reason, future industry activities in the gold market will need to be consistent with the market's price developments.

INVESTORS

It is quite hard to subsume investors in one group because there is so little homogeneity between these market entrants. It is nonetheless destined to play a major role in the gold market of the future.

Investors may be split into two categories depending on their investment decisions: long term and strategic investors. Although both pursue the same aims, i.e., generating a safe yield over the long term, individual decisions for or against such an investment widely differ.

Diversification, increasing the efficiency of a portfolio, inflation hedging, and generating an absolute yield are some of the core decisions concerning gold investments. The classic portfolio theory founded on the portfolio selection model proposed by Henry M. Markowitz tries to determine optimal portfolio composition. The aim is to find the right mix of asset classes for optimal portfolio efficiency. The efficiency is defined by the ratio of risk to yield, i.e. volatility to interest rate. The question of whether the admixture of gold provides any risk reduction has been explored by a number of specialists in the past.

I shall refrain from trying to answer this question empirically at this point. (A number of empirical studies, among them some studies very recently published by Yale University, have reached the conclusion that the admixture of gold in a depot does contribute to risk reduction and portfolio optimization.) I rather intend to show to what extent the gold market adjusts to these developments and how the "institutionalized investors" approach this market.

In a long term mean, gold correlates negatively to bonds and stocks. This property is of late being increasingly exploited by institutionalized investors to provide their portfolios with stability and balance. They frequently place only a small percentage of the total volume in the gold market, usually ranging around five to ten percent of the entire portfolio. The gold position can and is reflected via the various instruments explained above. Because these gold positions are of a long term nature and often serve the purpose of diversification, the opportunity costs are ideally kept to a minimum. To hold the gold in a metal account would be unsuitable in this case as the annual fees are too high. Gold certificates are much better suited, as they are less cost intensive and can be managed more flexibly.

Another motive for gold investments is the property of gold to correlate positively with rising inflation rates. In an inflationary environment, gold offers security and stability of value, as well as an emergency reserve, even when it is bought as a long-term portfolio investment. In insider circles the portfolio percentage thus invested is usually put at around five percent. But gold investors primarily motivated by these reasons do not usually opt for gold derivatives like the gold participation certificates common at the moment, but buy physical gold bars instead. The opportunity costs are usually neglected due to the significance of the physical purchase.

Another part of the portfolio approach is the absolute return idea. Gold suits the current Zeitgeist very well. The gold price is rising and is more volatile than it has been in years, and hence offers ideal conditions for profits in gold. Institutionalized investors thus increasingly demand leveraged derivatives and options in recent years in order to participate in the price market.

These developments can be easily proven, most simply by taking a look at the variety of gold products currently available in the market. An examination of the German market shows that virtu-ally all major issuers now offer gold structures via the exchange or OTC. All kinds of warrants, Asian option structures, knock out options, certificates, Quanto-products and turbo options have become available, whether listed in an exchange or as an OTC product, with or without foreign-exchange risk coverage. The turnover figures vouch for the heightened interest, and the market has also been experiencing a greater demand volume since the year 2002. Because one thing is clear: Each product—be it an option or a certificate—needs to be traded via the spot market by any issuer wishing to cover possible price change risks. An option sold in England or Germany therefore immediately creates a demand in the physical market.

The demand is highlighted even more if we take a look at the exchange traded fund product listed by the World Gold Council, the gold bullion securities. This product is a classic participation certificate backed by physical gold, if the World Gold Council is to be believed. It has already been listed in London and Sydney, and most recently in New York, where the demand is rumored to have reached over 100 tons of gold within the very first days. This gold disappears from the market and is transferred to investors' depots. And even if products of this type are frequently resold and bought, a large part remains as long term residue.

It is to be expected that most family offices and high-net-worth individuals will increase their gold market engagement in the future.
Because it is one of their express aims to primarily maintain family
and other assets; that is, protect them against value depreciation
in the shape of inflation. It is particularly the risk of inflation that
motivates investors. Their fears of massive inflation are not entirely
unfounded, particularly if one considers the large budget deficits
of the USA and some European states and, most of all, their high
leverage factors.

But despite these recent developments one shouldn't forget that
gold holdings in portfolios were also a standard during the 1970s
and 1980s. Gold used to be a staple of balanced asset allocation
globally. But with the price depreciation in the 1990s, gold was
all but banished from institutionalized investment portfolios. This
makes the most current trend toward including gold in portfolios
again less of a novelty, and certainly somewhat more comprehensible.

The year 2004 marked a 16-year high in the gold price. The result
for the gold market was a larger audience, which was able to
peruse a growing variety of products. It was not only the press and
the media that directed investor attention to the precious metal; the
investment committees of banks and large asset management firms
were just as busy promoting it.

The prognosis for the next few years hence seems to promise a
favorable gold market environment. The "old," and now rediscov-
ered, investment class gold not only increasingly makes a come-
back in the portfolios of institutionalized investors, but it is also
able to once again attract speculative interest. The II. Washington
Agreement on Gold has rendered the central banks much more
predictable for the next few years. All sales will be coordinated,
meaning that the pressure on the gold market will be minimized,
and sales can only have a positive effect on prices. The mining
industry is stagnating, and it will certainly take several years until
new major projects such as that of Ivanhoe Gold Mines, Qyu Tol-
goi in Mongolia, will take up production. If the investment environment remains positive, the gold price will continue its uptrend.

But where there is light there is also shade. For the gold market
this raises the question about what the potential risks of the positive trend reversing are.

One of these risks is the danger of strong gold price fluctuations, of course. But a second, far greater risk that could probably cause droves of asset managers to desert the gold market is that of a worldwide economic recovery, along with a collapse of geopolitical risks, a reduction of balance deficits and a curbing of leverage factors. The whole thing could be accompanied by market entrants distancing themselves from assumptions of rising in- or deflation. The economic stability of the great powers, stable currencies and a reduction in geopolitical risks, which dominated the agenda in 2004, would also be a world where gold would only find a place and vindication in private and institutional investors' depots with some difficulty.

But no matter which scenario is envisaged or desired, it will first have to deal with realities. And the fact is that the gold market has experienced such invigoration and gone through so many changes in recent years, one is tempted to assume the market is a "new market." But as you have witnessed, it has principally remained the "same old market."

My intention with this book was to bring greater transparency to one of the world's oldest financial markets. To this end, I have described the instruments and products used in the market, supplied a short historical overview of the various gold markets and taken you, worthy readers, right up into the present, including the trades now customary. This is by no means the conclusion of a look at the market: Details and specialist knowledge only familiar to active market players also deserve closer scrutiny. But I do hope that this book managed to provide a reasonable foundation for a deeper understanding of this richly traditional and interesting market.

Only the future can show what the gold market's further developments will be. But one thing is certain, even today: There will always be a market for this beautiful and valuable yellow metal, be it underground or publicly accessible.

APPENDIX A: WEIGHTS AND MEASURES

Weight Statement	Used in
Gram	Europe, but also in use internationally
Ounces	Anglo-Saxon countries: USA, UK und Australia
Tolas	India, Pakistan, Near East, Singapore
Taels	Chinese-speaking countries: China, Hong Kong, Taiwan
Baths	Thailand
Chi	Vietnam
Dons	Korea

Global Weight Measures in Precious Metal Market
Source: *NM Rothchild & Sons*

	Ounces	Gramm	Kilograms	Taels	Tolas
Ounces		31.1035	0.0311	0.8310	2.6667
Gram	0.0322		0.0010	0.0267	0.0858
Kilograms	32.1507	1.0000		26.7172	85.7550
Taels	1.2034	37.4290	0.0374		0.3212
Tolas	0.3750	11.6638	0.0117	3.1135	

Conversion factors of most common weight units
Quelle: *LBMA und eigene Berechnungen*

APPENDIX B: LBMA MEMBERS LIST

I. Market Makers

The Bank of Novia Scotia – ScotiaMocatta
(gold fixing member, general clearer)

Barclays Bank PLC
(gold fixing member)

Deutsche Bank AG
(gold fixing member, general clearer)

HSBC Bank USA, London Branch
(gold fixing member, general clearer)

J Aron & Company (UK)

JP Morgan Chase Bank
(general clearer)

Société Générale
(gold fixing member)

UBS AG
(general clearer)

APPENDIX C: INTERNATIONAL ASSOCIATIONS

Australia: AGR Matthey; Pasminco Metals Pty Ltd.
Canada: Royal Canadian Mint
China: Great Wall Gold & Silver Refinery of China
Germany: Baden-Württembergische Bank AG; Umicore AG &
    Co AG; WC Heraeus GmbH & Co. KG
Ghana: Ashanti Goldfields Company Ltd.
Japan: Mitsubishi Materials Company Ltd.; Sumitomo Metal
    Mining Co. Ltd.; Tanaka Kikinzoku Kogyo K.K.
Kasachstan: OJSC Kazzinc
Korea: Korea Zinc Co. Ltd.
Mexico: Met-Mex Penoles, SA de C.V.
Netherlands: Schöne Edelmetaal B.V.
Poland: KGHM Polska Miedz SA
Singapore: Mahesh & Co. Pte Ltd.
South Africa: AngloGold Ltd.; Harmony Gold Mining Co Ltd.;
    Rand Refinery Ltd.
Switzerland: Argor-Heraeus SA; Cendres & Métaux SA; Finorafa
    SA; Metalor Technologies SA; PAMP SA; Valcambi SA
Turkey: Goldas AS
United Arab Emirates: A.R.Y. Refinery; Bin Sabt Jewelry (LLC); Kaloti
    Jewelry Group; Lakhoo Jewelry Trading Co. (LLC); Transguard
Great Britain: Alex Stewart Assayers Ltd.; BSI Inspectorate Ltd.;
    GFMS Ltd.; JBR Recovery Ltd.; The Bullion Desk Ltd.
United States of America: Gerald Metals, Inc.; Sovereign Bank
Uzbekistan: Almalyk Mining; Navoi Mining & Metallurgical Combinat

APPENDIX D: INTERNET - A SELECTION OF LINKS TO RELATED CONTENT

www.kofex.com
www.heraeus.de
www.cbot.com
www.umicore.com
www.comex.com
www.miningmx.com
www.goldseiten.de
www.allgemeine-gold.de
www.thebulliondesk.com
www.perthmint.com.au
www.lbma.org.uk
www.randrefinery.co.za
www.goldfixing.com

www.musuku.com
www.tocom.jp
www.argor.com
www.gfms.com
www.valcambi.ch
www.gold.org
www.brinks.com
www.bmf.com
www.viamat.com
www.pamp.com
www.metalor.ch
www.mks.ch

GLOSSARY OF TERMS:
THE ABC OF GOLD OPTIONS

(Excerpt with specialist terms relevant for this book)

American Option:
An option that may be exercised during the period and, respectively, is freely marketable during the period. This type of option is also traded at the COMEX in New York.

Asian Option:
An option whose calculation is based on an intrinsic value's average price over a specified maturity period. Asian options are most common in the metal industry because these companies base their metal sales and purchases on monthly price averages.

At-the-Money:
This stage is reached by an option if the strike price equals the underlying's market price.

Black / Scholes:
The most commonly used mathematical model for the calculation of option prices (i.e. of option premiums).

Call Option:
A buying option entitling the buyer to purchase a specified underlying value (in this case gold) at a pre-determined price at a certain future point in time. The seller of these options is in turn obliged to deliver the underlying value (the gold) if the buyer wishes to take up his option.

Covered Option:
An option where the seller actually possesses the underlying in order to deliver it in case the option is exercised.

Delta Hedge:
Expresses the volume a hedge position in the underlying needs to have in order to safeguard against the probability of the option being exercised. The Delta for calls is always between 0 and 1 and that for puts between 0 and -1. The Delta of an out-of-the-money option will tend toward 0 and that of an in-the-money-option toward 1. An at-the-money-option usually has a Delta of 0.5.

European Option:
An option that is only exercised at maturity.

Expiry Date:
The date when an option is due (i.e. when it either expires or is exercised by the buyer).

In-the-Money:
This stage is reached by an option when the strike price is lower than the market price, bestowing an intrinsic value on the option.

Intrinsic Value:
An option has intrinsic value when it is in-the-money. The intrinsic value is the difference between strike price and the underlying value's market price.

Leverage:
Describes the relationship between capital employment and potential profit.

Min-max:
A strategy whereby a client sells a call to finance the simultaneous purchase of a put option or sells a put in order to finance a call. The difference between the call strike and put strike is called the "margin," while the strikes represent the maximum profit or loss, respectively.

Out-of-the-Money:
This stage is reached by an option when the strike price is higher than the market price and the option therefore only has a current value.

Over-the-Counter Option (OTC):
An option that is not traded as a standard but bilaterally negotiated. This is most commonly the case with gold options.

Premium:
The price to be paid for the purchase of an option. The premium is usually quoted in USD per ounce but sometimes also in volatility percentages. The premium is composed of the following components: implicit volatility, intrinsic value, current value, interest rates, strike and general market supply.

Put Option:
A seller's option. The buyer is entitled to sell a specified underlying value (in this case gold) at a pre-determined price at a certain future point in time. The seller of this option is in turn obliged to

receive the underlying value (the gold) if the buyer wishes to take up the option.

Strike :
Price at which the option is invoiced when it is exercised.

Time Value:
Describes the period between the sale and maturity of an option. Longer periods increase the probability that the option will be exercised, hence increasing the importance of the time value.

Volatility:
The key determinant of option price calculations. The historic volatility can be calculated from the gold price's standard deviation. It forms the basis of the implicit volatility which expresses the volatility anticipated in the gold market in the future.

SPECIALIST TERMS - ABBREVIATIONS - INSTITUTIONS

Absolute Return:
An investment aimed at absolute capital growth rather than a relative one.

Allocated / Unallocated:
Allocated gold is stored individually; unallocated gold is stored collectively.

Assayer:
Assayer is another term for refiner or refinery.

Backwardation:
The opposite of contango, i.e., an interest reduction. The term is also used in connection with futures and forward trades.

Bankable Feasibility Study:
This is the study a bank requires from mine operators in order to start financing a project. The financing is principally based on the study, which is created by independent companies.

Benchmark:
A term of performance measurement. Benchmarks are used to measure the portfolio's performance relative to a specific reference value, for example an index.

Biscuits:
Gold bars that are most commonly used in Asia. They have a rectangular shape.

Broker Pit:
The trading floor has several areas. Traders and brokers who are active on the floor are accommodated in so-called pits. These are small, separated areas that function as a type of office. Brokers have their PCs and telephones in the pits in order to contact clients and colleagues.

Contango:
An interest premium. The term is used synonymously for both forwards and futures. CTA (short for Commodity Trading Adviser)

Deep Drilling:
A term used to describe mining projects where the ore is located several thousand meters below ground. Deep drilling is the typical exploitation method of the South African mining industry.

Deferred Payment:
The term for a remittance for which a deferred payment date has been agreed on.

Doré Bar:
A gold bar in raw form. Mines and refineries produce these Doré bars for further processing. They contain proportionate amounts of other metals besides gold.

Flag (here used in its specialist meaning for fixings):
A term used in the LBMA fixing. During the fixing, members can gain a period of grace by not knocking down their Union Jack, one of which is placed on every fixing member's desk. The process is termed a flag. Although the fixing doesn't take place as a physical meeting any longer, the term flag is still in use.

Floating Gold Price:
A term dating from the period when the fixed price for gold ($35 USD) was abandoned. It describes a gold price with a free price development, without restraints or levies.

Floor, Ring:
Various parts of the trading floor. The floor is the actual floor where the deals are made. The London Metals Exchange's floor is also called a ring.

Floor Trading:
Only very few exchanges still trade gold on the floor. But it is still standard at the COMEX in New York.

General Clearer:
The services of a clearer are usually offered by banks. The clearer serves as a settlement, storage and management location for the gold.

Hedge Funds:
Funds trying to make a profit in various products. They also use products initially designed as safeguards, so-called hedge products.

Hedging, Hedges:
Hedging is the safeguarding against risks via finance contracts. Examples are sales for forward delivery or option trades.

Hub:
Gold hubs are cities or even regions that achieve large turnovers or, respectively, numerous business deals with gold in relative terms.

Lending:
Gold can be borrowed. The terms are the same as in purely financial transactions: lending and borrowing.

Limit Up/Limit Down:
A regulation in futures exchanges permitting only a certain fluctuation range per day. The upper and lower limits are called limit up and limit down, respectively.

Locals:
The term is used for futures exchange traders who act on their own account. They often adopt a dual role whereby they also execute trades on a commission basis.

Long/Short:
Two trading terms. Going long means buying something; going short means selling something.

GOLD

Margin (Initial/Maintenance):
Margins are deposits required for futures contract trades. In futures trades, only a portion of the total price traded is required as a credit security. This portion is called the margin.

Market Maker:
Markets depend on a certain number of traders and banks who are prepared to permanently quote two-way-prices, i.e. a buying and a selling price. Market makers undertake to continuously quote two-way-prices for certain minimum volumes over a specified period of time.

Mine Financing:
Financing of mine projects on a project base, credit base or also a synthetic loan base.

Notice Period:
Before a futures contract becomes due, there is a notification period during which the buyer gives notice as to whether he wishes the gold delivered or not. The same applies to sellers.

Nugget:
A type of gold typically found in surface layers. Australia is famous for its "nugget fields."

Open Outcry:
A term describing a certain method of trading on a trading floor. The brokers freely shout their demand and supply positions. Deals are initiated by eye contact or body language.

Open Pit Mines:
Mines of shallow depths, usually in a funnel shape.

Open Position:
The open position in a futures market shows how many unfilled positions there are in the market. Their number also gives a clue to the behavior of market entrants.

Outright:
A position containing an open risk is called an outright. If a trader sells a gold option without actually owning any gold the position is called an outright position.

Overnight Orders:
Trade orders that remain valid until the next morning.

Retail Investment:
A term for investments in the broader segment, i.e., in mass business. Products developed for retail clients are aimed at the quantity business.

Scrap:
Scrap gold is old gold recycled from the jewelry and electronics industries.

Small Bars:
A term used for all bar sizes below one kilogram, i.e., 10-TT-bars down to bars of a few grams only.

Spot (next):
A term for a trade's settlement date. Spot refers to two days after entering into the trade.

Spot Market:
Spot is the term for a settlement date in two days' time. The spot market is the classic inter-bank gold market. It is the basis of futures and options trades.

Spread:
The spread is the difference between the buying and selling price. One can also pursue a so-called spread strategy by speculating on increasing or decreasing differences.

Standard Transactions:
An expression used for physical gold trades at the Istanbul gold exchange.

T.T. Bars (Ten Tola Bars):
Tola is an Indian weight measure. Until very recently, the 10-TT-bar was the customary bar size in Indian trading.

Voice Dealing:
The gold trade still maintains its verbal form today in numerous places. The term voice dealing describes the manner in which deals are closed.

BIBLIOGRAPHY

Bähr, Johannes (1998): *Der Goldhandel der Dresdner Bank im Zweiten Weltkrieg*, 1st Edition, Gustav Kiepenheuer Verlag, Leipzig.

Barnard, Mark (1998): *"The Role of the Markets – New York,"* Director Mitsui & Co, speech at the FT World Gold Conference 1998, Barcelona.

Borgas, Fernando (1997) "Casting Brazil's Golden Rules," *The Alchemist*, published by the LBMA, number 9.

Botting, Douglas/ Sayer, Ian (1984): *Nazi Gold*, 3rd Edition, Mainstream Publishing, Edinburgh.

Capie, Jonathan / Mills, Steven/ Wood, Barbara (2004): *Gold as a Hedge Against the US-Dollar*, published by the World Gold Council, London.

Cross, Jessica (2000): *Gold Derivatives*, 1st Edition, Center for Public Policy Studies, London.

Desai, Ashok (1999): "India: Living with the Paradox of Liberalisation and Higher Import Taxes on Gold,: Indian Minister of Finance, speech at the FT World Gold Conference 1999, London.

Desebrock, Nigel/ Cross, Jessica (2002): *An Introduction to the Indian Gold Market*, Virtual Metals Research & Grendon International Research, Perth.

Desebrock, Nigel (1998): *Goldrefiners & Bars Worldwide*, 1st Edition, Grendon International Research, Perth.

Eibl, Christoph / Mezger, Markus (2003): "Gold und Silber: die langfristigen Aussichten," published in *Smart Investor*, Number 12/03.

Eibl, Christoph (2004): "Gold, Silver, Orange & Vanilla – The Colorful State of Commodity Investment in Germany," published in *The Alchemist*, Number 34.

Eibl, Christoph (2004): *Goldene Zeiten – Edelmetalle fürs Depot*, 1st Edition, published in Ralf

Vielhaber (Hrsg.): Was tun, wenn die Zinsen steigen, Dr. Theodor Gabler Verlag, Wiesbaden.

Endlich, Lisa (1999): *Goldman Sachs – The Culture of Success*, 1st Edition, Alfred A. Knopf, New York.

Fava, Peter (1999): "Change in the London Market and the Implications for the LBMA," Chairman of the LBMA, Speech at the FT World Gold Conference 1999, London.

Fava, Peter, (1998): "The Role of the Markets – London," Chairman of the LBMA, Speech at the FT World Gold Conference 1998, Barcelona.

GFMS (2004): *The Gold Market in China:* A New Beginning, published by Gold Fields Mineral Service, London.

GFMS (2004): *Gold Survey 2004*, published by Gold Fields Mineral Service, London.

GFMS (2003): *Gold Survey 2003*, published by Gold Fields Mineral Service, London.

Green, Timothy (1999): *The Millennium in Gold 1000 –1999*, 1st Edition, Rosendale Press, London.

Green, Timothy (1993): *The World of Gold*, 1st Edition, Rosendale Press, London.

Harmson, Stephen (1998): *Gold as a Store of Value*, published by the World Gold Council, London.

Ikemizu, Yuichi (1996): "The Tokyo Commodity Exchange," in *The Alchemist*, published by LBMA, Number 5.

Kleinmann, George (2001): Commodity Futures and Options, 1. Edition, Financial Times Prentice Hall, London.

Lips, Ferdinand (2001): Gold Wars – *The Battle against Sound Money as Seen From a Swiss Perspective*, 1st Edition, The Foundation of Monetary Education, New York.

Löhr, Andreas (1986): *Gold und Goldfutures*, 1st Edition, Duncker & Humblot, Berlin.

Mezger, Markus (2003): *Edelmetalle – Der neue Megatrend der Geldanlage?* 1st Edition, in: Anlegen am Wendepunkt, Dr. Theodor Gabler Verlag, Wiesbaden.

Mezger, Markus / Stahl, Markus / Single, Gerhard / Krauss, Michael (2000): *Gold – ein neuer Megatrend?* published by the Baden-Württembergische Bank AG, Stuttgart.

Mezger, Markus / Stahl, Markus / Eibl, Christoph (2003): *Megatrend Gold: Neue Entwicklungen*, published by the Baden-Württembergische Bank AG, Stuttgart.

Mezger, Markus (2004): *Megatrend Gold!* Phase III, published by the Baden-Württembergischen Bank AG, Stuttgart.

Murray, Stewart (1998): "The Changing Pattern of Physical Gold Flows throughout Asia 1997-1998," Chief Executive of the GFMS, Speech at the FT World Gold Conference 1998, Barcelona.

O'Callaghan, Gary (1993): *The Structure and Operation of the World Gold Market*, 1st Edition, The International Monetary Fund Publication, Washington.

Unnamed author (2000): *Gold Futures and Options*, New York Mercantile Exchange, Published by the New York Mercantile Exchange.

Unnamed author (2004): *The Gold Market in China: A New Beginning*, Published by the Gold Fields Mineral Service.

GOLD

Unnamed author (1999): *A Guide to Metals Hedging*, published by the New York Mercantile Exchange, New York.

Unnamed author (2003): *Introducing Gold*, published by World Gold Council, London.

Unnamed author (2000): The Companies Acts 1985 and 1989, Memorandum and Articles of Association of The London Bullion Market Association, London

Unnamed author (2001): The LBMA Precious Metals Conference 2001, Conference Proceedings, Istanbul.

Unnamed author (1999): *Introduction to the Gold Market*, published by NM Rothschild & Sons, London.

Unnamed author (2003): *Introducing Gold*, published by the World Gold Council, London.

Unnamed author (2003): *The Shanghai Gold Exchange*, published by the Marketing Committee of the Shanghai Gold Exchange, Shanghai.

Unnamed author (2004): *Trading Futures – An Introduction*, published by the Chicago Board of Trade, Chicago.

Unnamed author (2001): *A Guide to the London Bullion Market*, published by the London Bullion Market Association, London.

Unnamed author (2003): *Shanghai Gold Exchange*, published by the Shanghai Gold Exchange.

Pulvermacher, Katharine (2004): *Gold and Hedge Funds: a Comparative Analysis*, published by the World Gold Council.

Rockman, Ashley (2003): *Hedging in the Mining Industry*, special study, published by PriceWaterhouseCoopers, Melbourne.

Rosenthal, David / Young, Eric (1987): *Understanding the Gold Market*, 1st Edition, The Institute for the Preservation of Wealth, Bethel.

Sarnoff, Paul (1980): *Trading in Gold*, 2nd Edition, Simon & Schuster, New York.

Sitt, Robert (1995): *The Hong Kong Gold Market*, 1st Edition, Rosendale Press, London.

Smith, Russell (1998): "The Role of the Markets – Hong Kong and Tokyo," Director ScotiaMocatta at the FT World Gold Conference 1998, Barcelona.

Spall, Jonathan (1996): "The Asia / Pacific Market," in *The Alchemist*, published by LBMA, Number 5.

Tan, Ronald (1981): *The Gold Market*, 1st Edition, Singapore University Press, Singapur.

Warwick-Ching, Tony (1993): *The International Gold Trade*, 1st Edition, Woodhead Publishing Limited, Cambridge.

Whyte, James / Cumming, John (2004): *Mining Explained*, 1st Edition, published by The Northern Miner, Don Mills.

Zhe, Wang (2003): "China's New Gold Exchange," speech by the president of the Shanghai Gold Exchange at the 2nd Annual Gold Summit Conference, London.

ACKNOWLEDGEMENTS

Too many people to mention here have shown their support in the past. But some of them have become friends whose support is like a permanent fixture—not only in personal terms, but also in professional and technical areas. I would herewith like to express my gratitude to them, for their support, nurture and good company. A special debt of gratitude is owed to:

Christoph Metzger
Markus Mezger
Dr. Heiko Leschhorn
Dr. Markus Stahl
Brendan Downes

I would also like to thank the team at FinanzBuch Verlag for their fast and competent work; Mrs. Renate Oettinger for editing.
And a very special thank you goes to Ina Haug: Thank you very much for your unstinting care and support.

Christoph Eibl in Bretton Woods
Source: *Christoph Eibl (Photo)*

TRADING RESOURCE GUIDE

SUGGESTED READING

Forex Trading Using Intermarket Analysis: Discovering Hidden Market Relationships that Provide Early Clues for Price Direction
By Louis B. Mendelsohn

In today's global marketplace, currency values fluctuate every day and foreign exchange is the biggest market of them all, trading well over $1 trillion a day—more than all other markets combined! Master this market that never sleeps, and you could be the big winner. Just to survive in the hottest marketplace in the world, you will have to learn to stay one step ahead of the game.
This book is intended for traders and investors who use technology to win.

Item # BC109x 4183039 $19.95

Technical Analysis Simplified
By Clif Droke

Here's a concise, easy reading manual for learning and implementing this invaluable investment tool. The author, a well-known technician and editor of several technical analysis newsletters, distills the most essential elements of technical analysis into a brief, easy to read volume.

"A great primer covering all the technical analysis basics every active investor needs to know."
-Martin Pring, Martin Pring on Market Momentum

Item # BC109x 11087 $29.95

Bulletproof Your Trading Profits from IRS Attacks
By Ted Tesser

Join tax expert Ted Tesser in this comprehensive and illuminating presentation on what he describes as a voluntary U.S. tax system that exists for the informed as well as the ignorant. This disc intends to inform every viewer on the significant laws, rules, plans, and guidelines that will aid those who wish to pay all the tax that the law demands… but not a penny more.

DVD Item # BC109x 4120040 $99.00

The New Trader's Tax Solution: Money-Saving Strategies for the Serious Investor, 2nd Edition Updated
By Ted Tesser

Keep the profits you work so hard to earn and take control of your financial destiny with this thoroughly updated guide to reducing tax liability for the serious trader and investor. Jam-packed with invaluable business, estate, retirement planning, and tax-saving strategies that virtually anyone can implement within the new tax laws, CPA and expert tax consultant Ted Tesser provides current solutions for the tax problems most U.S. traders, investors and income earners are facing today. Comprehensive and simple to use, this complete guide includes not only proven methods to reduce your tax exposure, it comes with real-world case studies, illustrations, templates, and filled-out, ready to be filed tax forms. Whether you are looking for last minute tax-saving tips or trading techniques that position you for the optimal tax reduction, *The New Trader's Tax Solution* is a must-have to any trading library.

Item # BC109x 3252041 $59.95

The Profitable Trading Attitude
By Toni Turner

Big profits are made from "dumb" money or trades where the person on the other side of the trade is not smart enough to see what they are giving away or, more likely, too emotionally involved to let go.

Toni Turner now provides a step-by-step guide to make sure you are on the winning side of those trades. By developing a trading plan and visualizing those trades, you'll be able to:

- Establish the tone of your day to provide the best possible chance for gains
- Create a routine that makes the most critical element of your day an unbreakable habit
- Survey the market climate to determine what trades will win
- Set up your trades to achieve your goals in any market climate

This breakthrough course will provide you with a personal evaluation system that exposes the emotional baggage that is costing you money in every trade. It will then break down these obstacles and hardwire your core strengths to your trading plan to significantly improve your percentage of profitable trades.

Item # BC109x 3521305 $499

**To order any item listed
Go to www.traderslibrary.com
or Call 1-800-272-2855 ext. 109**

Free 2 Week Trial Offer for U.S. Residents From Investor's Business Daily:

INVESTOR'S BUSINESS DAILY will provide you with the facts, figures, and objective news analysis you need to succeed.

Investor's Business Daily is formatted for a quick and concise read to help you make informed and profitable decisions.

To take advantage of this free 2 week trial offer,
e-mail us at customerservice@traderslibrary.com
or visit our website at www.traderslibrary.com where
you find other free offers as well.

You can also reach us by calling 1-800-272-2855
or fax us at 410-964-0027.